BEYOND

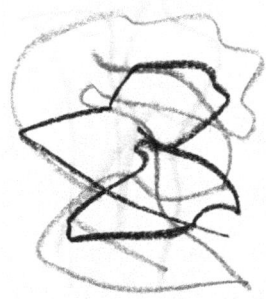

Light and Dark

Michael J. Osborne

Plain View Press
P. O. 42255
Austin, TX 78704

plainviewpress.net
sb@plainviewpress.net
1-512-441-2452

Copyright © Michael J. Osborne, 2009. All rights reserved.

ISBN: 978-0-9819731-9-7

Library of Congress Number: 2009931960

Other works by Michael J. Osborne

Spectra (Spectra Media)

Lease the Wind (Spectra Media)

Silver Haired Daddy (JF&M Media)

Lightland, Climate Change and Human Potential

(Plain View Press, 2000)

Silver in the Mine

(Austin Energy Publishing, 2002)

The Day of the Heart

(Plain View Press, 2005)

earthfamilyalpha@blogspot.com, 2004 to present)

Beyond Light and Dark

Contents

Book One   7

Introduction	9
Doors	11
Fields	27
Sky	41
Naming	57
Seeing	75
Being	87
Believing	101
Epilogue	115

Book Two   117

Introduction	119
The Wind	121
The Sea	135
The Creation	153
Dreaming	173
Working	189
The Heart	207
Beyond	223

BOOK ONE

Introduction

This book is not about religion. It is not intended to conflict with your belief in God. I believe in God. Intently. It is not meant to make you question your belief system in which Jesus is God. It is not meant to make you question your belief that Jesus is not the Messiah and that God is Jehovah. It is not meant to make you question your undying (or dying, as the case may be) obedience to Allah. Your faith in Buddha will not be undermined. Your love of the Cosmic Christ will not be diminished. Your understanding of the Tao will not be eroded. Your love of humankind and all of creation will not be questioned.

Whether you are a Zen Buddhist or a Full Gospel tongue talking Pentecostal lover of Jesus, you will not find this doctrine to be antagonistic. This is not the way of the Brujo or the message of the mage. These beliefs in consciousness of the spirit soul are not to be affected by the words in this book.

Rather, this book is more about learning to read. Or to count. This is not mysticism. This is learning. This is where we are. It is our place and our destiny.

This book is about understanding and applying that understanding as we participate in the world of space and time. It is not about moving towards the Light and away from Darkness. It is not about good and evil. It is not about the Archangel Michael slaying the dragon, that great Satan. It is not about leaving Hell and going to Heaven. It is not about the two lands that are in festivity. It is not about incarnation, astrology, phrenology, or pharmacology. It is not about the end of the World or the return of the Christ or the Resurrection of the Saints.

It is about the end of your world. It is about the end of a consciousness that has served us well and now, like a good plow horse, must be put out to pasture. It is a consciousness that is beyond light and dark.

DOORS

I have some really good news and some bad news.

The good news is this. You are reading the words of an enlightened being. I am. Really.

The bad news is this. Enlightenment is not all it's cracked up to be. And, this is one version of what it looks like. Now, before you go all negative or reactive on me, and stop reading, or even worse, continue reading with that big opinion you just made, let me assure you that you are probably an enlightened being too.

Enlightenment may not be a state. It is more likely a process. It is a moment by moment moment. Kind of hard to get I guess, but there you have it. Anyway, we all do it. We all know how to be there. A really good football game is a good example. In fact, the Super Bowl is playing today. There will be several hundred million people watching 22 men on a field of green throwing and running with an odd shaped ellipsoidal leather balloon.

From moment to moment, perhaps 200 million people will be totally focused on the play. The balloon will be set on the field. The team with possession of the balloon will meet as a committee behind the invisible line running perpendicular to the length of the field and come up with a plan. The team without the balloon will meet briefly after receiving signals from the sidelines where a highly paid assistant coach will make his best guess as to what the team with the balloon is getting ready to do to advance the balloon further down the field, ultimately across a goal which is called the zero yard line. Getting the balloon down to the zero is the way you score points. You get six when you do that and another one or two after that, depending on whether you kick the balloon or try to cross the zero line one more time from the three yard line.

This process is so engaging that millions and millions of people can watch the process and totally lose themselves in the moment. They find the infinite moment time and time again. The center hikes the ball to the quarterback who has his hands in the 300 pound 6 foot five inch bulked up Swede's groin. The quarterback, who is generally the leader of the committee with the balloon, gives the balloon over to another larger, generally stronger, black human who runs mightily in a crash of muscle and pads towards the zero line only to be all too often knocked to the ground where the balloon is placed at the site of the collision whereupon the next committee meeting is initiated and the process continues. (There are rules of course)

During this process, hundreds of millions of humans will hold their breaths in joyous participation in a process that takes their consciousness and gives it focus. It is consciousness candy. In that moment, we are one.

Good movies are the same. We love to lose ourselves in the 99 minutes of drama and excitement that is reflected into our eyes from the projection of light and images from behind our heads. The reflected images in front of our eyes and the surround sound THX provides ample conditions for our consciousness to be immersed in the lightshow before us. If we are not, it was not a very good movie.

We go from one event to another feeding our need to be one with the moment. We might lose our self in a good book. We might go to a basketball game. We might even watch TV. Whatever it takes to be one. We crave it. It feels good.

So you see, it is our nature to be transcendent.

It is our nature to seek oneness in the moment. We have developed as a culture an enormous industry to provide the consciousness candy we crave. There is the sports industry, the movie industry, the boating industry, radio and television, the music industry, the travel industry, the food and beverage industry, the drug industry legal and otherwise, and on and on. In fact, only 2 % of our industries actually provide us food and shelter. Much of the rest is consciousness candy.

And why do we crave this candy?

Because we are enlightened.

Now stay with me now.

If you do, you will see a door, and if you knock, it might open.

The Teaching

Go to the bookstore and find the section on spiritual development. (You don't have to go really, do it in your head) Many of the teachings in the new age movements teach that the door to oneness is found in the moment, the infinite moment.

Thirty years ago, the big popular book was Ram Das's "Be Here Now." It was almost the Bible of the Hip culture in the early seventies. The more cynical of us would make jokes about it saying "Be Over There Now." Caddy Shack said it best when the sage golfer (Bill Murray) gave the wise council to "Be The Ball."

Read anybody. It's all the same. Be in the present.

The Lord said in Matthew in the Sixth Chapter, verse 34, "Be therefore not anxious about tomorrow; for tomorrow will be anxious for the things of itself. Sufficient unto the day is its own evil. Seek yea first the kingdom of God."

Don Miguel Ruiz, the contemporary Toltec Master, tells us to be like the Jaguar. He teaches the art of stalking and of being totally in the present.

The present is the present. It is the Gift.

Human-kind has evolved enormously in the last 2000 years. Two thousand years ago, there were the great cultures. There were the Egyptians, the Chinese, the Persians, the Greeks, the Israelites. But few people could read. Even fewer could think. Fewer still could play an instrument. There were a handful of libraries and hundreds of manuscripts. The priests knew the secrets. They knew when the Sun would rise. They knew when it would return. And, they knew when the Moon would wax to its fullness and when it would wane back into the sun. Then it was magic and great knowledge. Today, it is in the fishing almanac.

There were great societies that were secret. Initiates of these societies were given great power by virtue of the secrets that were passed on through them. There was the power of the spoken word. There was the power of the subconscious. There was the power of the group, the power of song, the power of sacred prayer, the power of symbols and amulets, the power of crystals, the power of drugs, and the power of ritual and rite.

Today, we pledge allegiance to the flag. We sing the Star Bangled Banner before we play our balloon game. We drink Bud because that's "the kind of man I am." We live in a gated community because we are different and special. We read Harry Potter because everyone loves magic.

Almost everyone reads in the developed world. And that's more than half of the 6 billion humans on earth. A sixth grader has more knowledge of the earth sciences than an educated gentleman of the 1600's. A publicly schooled 15 year old looks farther into the depths of space and deeper into the cells of life than Di Vinci could have hoped.

As uncivilized as it all seems, we are really quite civilized. Most of us do not get hit over the head as we walk to gather berries in the bush. True, we may get hit by an angry motorist who is late for work or unhappy with his wife. As unsafe and weird as our cities are, most of us go to work and come home and nothing really happens. We go to work, answer our e-mail, check our Microsoft Outlook calendar, return our voice mail, and go about our day. We worry about the economy, whether or not we will lose our job, and how we are going to send the children to college.

We come home and eat radiated food that is really pretty safe, and we drink water with relatively modest amounts of arsenic and fecal matter in it. Our homes are generally warm in the winter and cool in the summer. Even the dreaded double wide trailer is a pretty nice space to hang your hat in when you compare it to the breezy, dusty shack of not that long ago. And if you go to the developing and undeveloped world, "not that long ago" is right now.

True, the double wide is soulless. True, McDonalds sells meat that has traveled to Timbuktu and back, and true, our cities are ridiculous. Yes, TV is for imbeciles. Yes, our Capitalists are greedy. Yes, our city governments are corrupt. Yes, we are killing the earth with our pollution. Yes, we are changing the climate with our burning of carbon. Yes, we are not free. Yes, democracy is really a plutocracy. Yes, things could be better. Much, much much better.

But, in fact, we have come a long way.

We have come so far, that many of us-let's be honest, are bored. We don't have to defend our family from the lion in the bush anymore. (Not that Bush). We will not be attacked by roving bands of thieves (unless we live in tough part of any city). We can, in short, be lazy. That's not a sin mind you, it just makes it hard to be at your best and it definitely makes for a "be some where else" state of consciousness.

"Oh if I can only get to the weekend." "I can't wait for the super bowl." "I can't wait till the play." "Oh, if I can just make it to Thursday when I will see my lover." "Oh I can't wait till we take that vacation and I don't have to come to work at least for a week." You get the idea. You probably even get it better than that.

So, in order to find our oneness, we watch movies and football games, and we drink beer and we smoke. Then one day, we find ourselves at death's door wondering how in the heck we got there? Our life suddenly seems to have become, well, meaningless. And let's be honest. It probably was.

"Oh, if I had just lived every minute as if it was my last." The secret to opening the door is exactly that. Live each moment fully. Stay in that place where you are most alive, not where you are most comfortable.

Now, I am an expert on this. Not at being in the moment, but at being in that safe place. (Ask the two wives that lived with me). But really, why do we seek this safe place?

Fear.

Fear keeps us from living. Fear keeps us in our cave of self doubt. Fear keeps us from living each moment as if it is our last. Fear keeps us from seeing the birds that fly and the eagle that soars. It robs us of our ability to see beauty. It is the burglar that comes each night and each day. It is the robber of your fortune and the thief of your fate.

Miguel Ruiz says that the Angel of Death should become your ally. If the Angel is not your ally, then you are dead to the moment anyway and death will surround you.

The great religions have made much use of this fear. Knowing that we are all subconsciously fearful of our mortal end, various religions have provided for their believers a mechanism to put this subconscious fear into some manageable psychic space. You do not need to fear death if you are going to heaven. You will go to heaven if you have been blessed by the priest. You will go to heaven because you have accepted Jesus as your lord and savior. You will go to heaven because you have lived according to the

Holy Koran. You will go to heaven and be judged for your works, so be a good man. You will go to heaven if you have confessed. Even the atheist has his spin. You do not need to fear death because there is no hell, only dissolution. In the east, there is no death, only reincarnation. The yogis would have you re- identify with something that can't die. Identify with light some say. You are light. O Death where is thy sting?

Or you can identify with Consciousness itself. It is not only light, but organized light.

It is the Word.

Or you can identify with life. Or your ego can identify with the all of the visible cosmos. You are God. You are the creation. You are immutable. You are eternal.

The Truth is. You are going to die. At least the YOU that thinks it is reading this book. Even the realized YOU is going to die. You are going to walk through the door of the world into the door of another world. You will die to this world.

Living in the moment is living in the door of death. It is taking your special place in the special place and time you have come to inhabit. It is your job to do it well.

Relax.

Play like you are at a great football game and the team with the balloon is about to score. Stay focused. Stay in the game. Get a beer. And when you do get that beer. Drink the beer.

When you eat... eat.

When you love, love.

When you listen, listen.

When you talk, talk.

When you watch, watch.

When you hurt, hurt.

Don't hide in your cave. Be alive and fresh and free. It's a tall order.

We all love to hang around people who are alive and fresh and free.

We wish we could be like them.

We can.

Next time you are introduced to someone, you can practice. When the name is given to you, let go of the wall that is keeping you from hearing their name and repeat their name to them. Keep repeating their name, asking them how they spell their name. Tell them you know someone who has that name and you do or don't look anything like them. Ask them who gave them the name. Come out of your cave and greet the person who has come to your door.

Stand at your door of consciousness and welcome them. Welcome them to the oneness. Welcome them to the oneness that resides in us all. You are the door as are they.

From time to time, I actually do this. The more I practice, the better I get. The better I get, the better it gets.

You are the fresh breath in the open door.

The Practice

The practice here is not to be loving or thoughtful or selfless. That will come later. The practice here is to be simply here. What follows are examples that presented themselves as I wrote these words on a Sunday afternoon.

I must stop writing now, because a friend of mine, cousin Artie Osborn, (he is not really a cousin, we just share the same last name except for the "e" on the end) has lost his business partner and long time friend. It was quite a passing. Doug, who has always been a bit of a legend, fell to his mortal end in an old dump truck on a narrow high mountain road in Honduras. The phone rings though, and it is Susan, a woman that I knew 33 years ago, and now is back in my life. I am going to go pick her up.

We go see a movie. It is Billy Bob Thornton's Monster's Ball. Now, that is a good movie. The love scene is great, and who doesn't want to feel better. But the end is masterful. Something in her face when she decides that it is going to be alright. Then, Billy Bob says "Honey, we're going to be alright." The camera moves out to the stars and the Cosmos.

The turning point for Billy Bob earlier in the movie is not really known, because it is saved for that moment of eternity.

It is Saturday. The wind is blowing but the sky is now clear. It is very cold for an Austin, Texas March 2nd. It is Texas Independence Day. I try to run Bookshelf to get some poetry or great works on Doors and it won't open. Can't read Drive D.

So, back to the practice.

Every moment is precious. Every encounter is full of infinite possibilities. Every second we participate in this great realm of consciousness is a blessing. Granted, it is very, very hard to get there when you have a headache or a heart ache, or even when you feel great.

What ever you do. Do it with consciousness. Cook your dinner thoughtfully. Wait at the red light thoughtfully. Find that place where the magic begins to take over. Because it will, and you will know it when it comes.

Practice, as the Buddhist say, "Mindfulness."

Practice when you pay the sweet man in the box who is collecting money for the multinational corporation who owns the parking garage at close to minimum rage.

Practice when you order your food. For really good practice, be as "good as you can be" after waiting in line at the Post Office to pick up some kind of awful certified mail for no good reason. "Love your enemies. Bless them that curse you. Do good to them that hate you and pray for them who despitefully use you and persecute you."

"For if you love them who love you what reward have you? Do not the tax collectors do the same? And if you greet your brethren only, what do you more than others? Do not even the (re)publicans do the same?"

"Be ye therefore perfect, even as your father in heaven is perfect."

Knock and the Door will open.

Practice being Here.

Welcome.

You will fail. We all fail. You will fail miserably. We all do. You will get disappointed. Then, you will be looking out the window and you will see some birds hopping around on the ground eating god knows what. They will chirp and hop and two more will join them. The wind will move the vine on the fence and a seed pod will drop to the concrete. You will suddenly see the perfection, the beauty and breath of it all.

It's like that sack that floated around like so much other trash in American Beauty. What is mundane becomes magic. Because, you see, it is magic.

Find that space and breathe with it. Don't think too hard; just be there. Find the tranquility and the peace. Listen to the wind. Feel the beat of the City. Hear the sirens in the city and the grumble of the highway. Hear the rattle of weighted windows and the chimes on the porch in the apartment across the way. Settle in to the moment. Unleash your potential. Don't think or even plan, just be. Be a human, being.

The moment will stretch like a rubberband. It will grow larger and longer and fuller and more pregnant until you feel like you have taken some kind of mind drug. You will not be anxious. You will not be angry with your boss or your spouse.

Imagine the earth turning towards the east. Feel the turning. Know that the moon is moving around the earth but very slowly. Know that you are on this big ball of condensed energy being afforded a complete view of the entire creation every twenty four hours, if you would only look.

Feel a tingle come up your spine. Move your muscles and give it a channel. Let it settle in the valleys of your muscles and the mountains of your mind. Don't think about it. Feel it.

Relax your face. Feel the tightness dissolve. Forget about the laundry for now. Listen.

Watch.

Become Still.

Become Alive.

See that everything can be seen from some other perspective. See the tree limb show itself on the shadow of the white stucco wall.

Breathe deeply in your stomach. Push your stomach out and pull the air into your cavity. Stop reading. Listen to the plane overhead. Listen as it gently fades into sounds of the urban cacophony.

Take a walk with your lover. Walk by the creek with your son. Look into the soul of a stranger.

– Beyond –

Be grateful to God or the Great Spirit or even your father and ancestors who have brought you to this moment. Breathe the air that we all breathe and know it fully for that moment. Take of the water that lets you live, knowing it is only a gift that will ultimately be returned. Feel the lightness in your being. Feel your heat. Feel the blood moving through your beingness.

On the Beach

Sitting, gazing on a moonlight eve,
my eyes found light
from the still waters below.
Resounding in my moment
the age old became new as now
and made its presence known.
Like Baba some one
it landed like a space plane
onto the runways of my being
and the plane of my place,
Its truth no longer lost
in the melodies of the mind.
It washed ashore
like a pier with no ground.
Dodging, weaving
daylight carves its way into meaning
and warms the memory
when Time found slumber
and the Moment awoke

FIELDS

We live in Fields.

We think in fields of knowledge, fields of words, fields of concepts, and Platonic Archetypes.

You cannot find Oak trees in a field of strawberries.

The small infant looks at the movement of lights and darks. It is a light show. These variations of light and shadow slowly but surely turn into shapes. There is the shape of the face of the one who holds me and sings to me and feeds me. Her voice is soft and her touch is warm. I know her. Direct Reality, so to speak, begins to turn into something else. The brain is working overtime trying to make sense of it all. Soon, there are other faces. There is this gruff dada who holds me in a different way. I think I have a sister who, like my mother, has a soft voice, but she is not so nurturing.

I am in a mystery. I have no idea what is happening to me.

Whoever me is.

I feel hunger. So I cry.

I am fed. So I sleep.

I wake up. So I cry. On my good days, I wake up and listen to the funny noises coming from the light box on the wall.

Sometimes it is very bright. Sometimes it is very dark. I cannot see in the dark. It makes me afraid. Whatever that is. Maybe, I'm afraid I won't get fed or get touched or receive the soft sounds that make me feel so good, whatever feeling good means. Feeling good must be the opposite of feeling bad. Good and Bad, Light and Dark, Hard and Soft. The reality is becoming more understandable.

My fields are expanding. When the mother form comes into my view, I can abstract that presence to mean that I will get fed or at least get this wet weird stuff taken away from whatever that is below me that seems to be attached to me. It smells... whatever smelling means. If it smells too long, it hurts. I know what hurt means. It means discomfort. It hurts dammit. Now I can cuss. Taking the Lord's name in vain is just around the corner.

There is day and there is night. I go to a soft place where I lay down and I disappear when it is night. If I stop disappearing, I cry. Often, a soft thing is stuffed into my mouth and I am fed. My belly feels warm. I am happy, whatever that means and I disappear again.

I don't like my brother. He doesn't like me either. I can feel it. Don't ask me how, I just can. I think there is something called jealousy that I will soon learn about. And there is this thing called attention which I really crave. Attention gets you anything. No attention is bad. However, there is also bad attention. I have not figured that out yet.

I am learning that there is an outside and an inside. Not just outside me and inside me, but also outside of the big us. Sometimes we get into another inside and move through the big outside. It makes me sleep. I don't understand that either.

It can be hot or it can be cold. I don't like cold, but I do like the soft thing next to me that make me warm. I am learning to use my breath to make noise. I can make the things on my face where my comfort comes into my body move when I force my breath. I make sounds. When I say something like Mama, I get held closely and warmly.

The mind grows, we learn to talk, and we learn to listen. We learn what our name is, and who we are, and who our family is. We learn where we live, what day it is, and how to put popcorn in the microwave. We build up a whole system of language and thought. The "Field" that at one time was undifferentiated and whole, is now cut and parceled into so many pieces and bits of reality. We have become the process of becoming intelligent.

But this intelligence comes at a cost.

The once Holy Field of Reality becomes only so many pieces of a dead universe. The magic of time and space and the Reality of Being becomes so many pictures. It becomes only what we choose to see. The Holy Field becomes only a field of objects to be manipulated and negotiated. It is as if we learn only to become drivers of cars. The Magic of Time is lost. The magic of the Universe falls behind a screen of ideas and constructs that in fact blind us from the original truth we were born into.

But this is no curse you argue. What we lose is more than offset by what knowledge we gain. Were it not for this ability of our brain and our minds to take this reality and break it into an understandable universe we would be no different from the lowest of creatures. Who would argue that our minds and our brains are our most valued of assets in the inventory of the corpus.

With our language and our minds, we craft tools and highways and homes and cereal named Fruit Loops. Our field of dreams slowly becomes a field of nightmares.

How do we see beyond this veneer? Or why would we even care to see behind the curtain again?

Is it not enough to drink good wine, watch good movies, and enjoy square cheeseburgers?

Perhaps so.

But can there be more?

Can our consciousness be attuned to even higher and more useful levels of perceiving?

Surely, our present level of perception is not the end of evolution or the final stage of the development of our thinking power. We have not conquered poverty or hunger. There is still much war, and there is much in our lives that could be improved. Many of us in the West forget about the Developing World and the pervasive poverty it contains. Certainly, we have enough on our own tables to say grace over. We worry about our jobs, our 401k's, and where we will get the money to send our children to the best colleges.

We reason that our consciousness is good enough to get them into school and hopefully, good enough to allow us to buy that house in that gated community. We reason that David Letterman is a decent enough way to spend the late evenings. And, in fact, it is.

These are our Fields. We live in this field of words and objects and jobs and cars and children and taxes. We fear loss. We fear the loss of our job, of our loved ones, of our sight, or our minds. We fear that someone we love will be transformed in the blink of an eye into a being who no longer can walk without a cane or move about without a wheelchair. We fear being robbed. We fear being raped, literally or financially.

And why should we not fear such things? These things are all about us. Only a fool or an idiot would not see such things. This world is a world of hard lumps and then you die. Yes. But, it is also a World of Perfect Chance and Magical Moments. How do we reconcile this dichotomy? We feel it when we see beautiful art or hear poetry. We sense it when we see movie fables of unlikely heroes like Forest Gump. We see the magic in the American Beauty plastic convenience store bag that floats like an angel against the backdrop of a brick wall.

We feel it when we go fishing and nature speaks its splendor to us. We feel it when we find ourselves looking into the eyes of our lover and best friend knowing that somehow we found each other and our two souls became one in Holy Marriage before God and men. We know we have more than just a bunch of molecules banging into each other. Yet, we are not really sure.

– Fields –

We might believe in God or Fate or Karma or Justice. We might feel that evil-doers will meet their maker someday. We may be certain that they are going to Hell, but we are not sure if we are really going to Heaven.

As the saying goes, "Everybody wants to go to Heaven, but Nobody wants to Die."

Our consciousness resides in these fields. It can and does also reside in the Holy Field. In the small field, a bicycle is given to your son for his 10th birthday. It is a good thing. He is happy. He makes friends and they ride around the neighborhood having good and loving communion with life and the reality they know. Then one day, a 16-year-old, high on glue, hits your son with his dad's stolen car and knocks him to the ground. Your son breaks his collar bone and both legs. This is a bad thing. The 16-year-old knows it is a bad thing. He is grounded from driving the car. Your son cannot go to school.

While grounded, the 16-year-old discovers that he likes physics. Someday he will discover antigravity. He will contribute substantially to the advancement of human kind. This is a good thing. Your son misses his ride to school because he is bedridden.

A nail falls out of the weekend carpenter's truck and punctures the rubber in the school-bound car. The flat causes the car to swerve into a truck driven by man who never hurt anyone or anything in his life. Everyone in the car dies. Your son is spared. That is a good thing.

The truck driver is stricken with grief. The weekend carpenter is a Pentecostal preacher. He is not aware of his part in this tragedy as he preaches on Sunday about the Lord working in mysterious ways.

The truck driver is so stricken that he becomes a drunk. He loses his home and his wife leaves him. The home is sold to a young black couple that has been looking for a good place to raise their three children for a year.

Your son, who is spared in the wreck, becomes a great basketball player. That is a good thing. He becomes a really good basketball player. That is a great thing. In the last 2 seconds of the Final Four Championship, he misses a free throw that gives the championship to the coach who miraculously wins his first National Championship. Your son is devastated. The retiring coach is lionized. You lose you position as the father of the hero and instead become the father of the guy who blew the National Championship. All of this because you bought your son that bike when he was ten.

This is a dead world of fields?

I think not.

Our consciousness is not really equipped as currently wired to explain this. Only in poetry and other communications of transcendence can it even be brushed against. The best we can say, "Wow, wasn't that weird?"

And, it is weird.

"Weird with a beard" as we used to say. What is good turns bad, what is bad turns good, what seems insignificant is crucial, what is important is nothing. Our description of the field turns out to not be a very good description at all. In truth, it becomes more of a lie that it is the Truth. And so it is.

Someone once said, "you have to lie to illustrate the truth." The Truth is out there, as the TV series says, but getting at it is a source of endless adventure.

Our description of the Field is just that. It is a lie. But it is a pretty decent illustration of the truth. But, it is only an illustration. Illustrations are pretty good descriptions of the scene, but that is all. A beautiful illustration of a horse will "not a ride make." A great colored illustration of the Live Oak tree on the corner next to the courthouse will not cool you in the summer.

– Fields –

A religious tract may point you to the Truth, but it will not BE the Truth. Our field is just an illustration of the Field it attempts to describe. Problem is, we forget that it is a lie. We get confused and think we are seeing the Truth when we are seeing a lie. We forget that the Holy Field is behind that curtain of words and mindforms and human understanding. When we do, we lose ourselves in that matrix. We lose ourselves in the Lie.

There is a wonderful Professor in Colorado who makes a speech about good things and bad things. He has a big long list. War is bad, but it reduces population which is good (in a scarce world). Death is bad, but it makes more food and makes for less pollution. War generally creates better medicines and medical treatments. That is good. Good health is good but that makes people live longer which makes more people and that is bad in an overpopulated world. Car wrecks are bad, but they... well you get the picture.

What is good is bad. What is bad is good. The Allies won World War II and did so by building a military industrial complex that could defeat the military industrial complex of Nazi Germany. To stop the Nazis we firebombed Dresden. To win the War against Japan, the US destroyed dozens of cities and murdered hundreds of thousands of their inhabitants. We did it with conventional firebombs and finally with the very first atomic bombs. Now, the US provides 60 % of the arms to the rest of the World. That is Bad. Power plants are good things. But pollution from their smokestacks kill 50,000 people a year in the US alone.

Cars are good. They kill 50,000 a year too. They injure several hundred thousand more. This is just in the United States.

This is our field. How do we make sense of it? Where does it stop? How do we judge?

The Lord said Don't.

The Teaching

Judge Not that Yea not be Judged.

For with what judgment you judge, you will be judged. And with what measure you use, it will be measured against you.

First, cast the beam out of thine own eye.

How would we judge the preacher who let the nail fall out of his truck that caused the flat, that made the car swerve, that killed the children, that ruined the life, that provided a home?

How do we even judge Judas who betrayed the Lord. It is said that the idea to betray him was planted by the Lord himself. Someone had to do it. Judas hung himself. Peter denied the Lord like a coward and yet now, he is the Rock of that Great Church.

Go Figure.

Judas, you could argue, did the heavy lifting that night in the Garden. He, by some standards, was more of a hero than Peter. Try starting a Church in his Name. There is no Saint Judas.

We judge him as a traitor.

But the teaching is to not judge, "And to remove the beam in your own eye." We, like the eagle, can observe though. From a distance, we can see from as large a view as our position and senses allow. We can see and we can discriminate, but we should not judge. For we cannot know who makes the rain. We cannot call the Wind. We cannot make our own Hearts Holy.

We can see and we can be thoughtful.

This is not an easy thing. It is much easier to see the small field of the nightmare and judge our self, our friends, our leaders, those who have harmed us, and strangers who cross our field of consciousness. We can judge them as stupid, or evil, or careless, or thoughtless. We cannot know their hearts, their cares, or their minds, yet we can judge them. We judge ourselves with the same vengeance. Yet Vengeance is not ours. When we do judge, we create the victim. And when the victim becomes judged, whether it be ourselves, our neighbor, or our mate, we create a field of losers and winners.

Within this field, we then create our own drama and our own fiction of the Truth. Knowing nothing we are "all knowing." Seeing only the veneer, we rage war on the injustice of the moment knowing nothing of the Truth that resides under every energetic moment.

We become angry. For we have been insulted by the man who cut in front of us. We are not going to be treated in this matter. I demand that you seat us now. We have reservations. They do not. I know what is right. You do not. I have made my judgment and it is right, and it is noble, and it is the truth.

Or, I will become angry. Or, I will send my business somewhere else. Or, I will file a lawsuit. I will send my army and my air-force. I will put a bomb on the backs of children and have them walk into your pizza parlour and murder all around them. I will show you that justice will prevail; that evil will not be tolerated.

I will show you that killing people is wrong. And, I will kill your people to get the point across. Yet we know that "an eye for an eye and a tooth for a tooth" leads to a blind and hungry populace.

The Lord says, "To resist not Evil."

We are told to Love Our Enemies. and to Bless Them that Curse Us.

This is not silly do gooderism. This is hard, deep understanding. This is a profound deep understanding of the nature and finite potential of the present field of consciousness that we employ.

This is not something we should do. It is something we must do. It should and must become our deep understanding of the limited value of the field of words and symbols that we employ to make sense of the Holy Field. How can we place such merit on shadows inside a cave?

The teaching on Fields then is this. Don't make so much of it.

There is Zen teaching that says, "There is so much blame in the World, so don't worry so much about it." The master of this saying knew of our limitations of perceiving the Truth.

Yes, the light is red for someone. Someone ran the light that dented your tranquility. So What? With What Judgment Shall YOU Measure? Honor Not That, That is Measured.

The Practice

It is not so easy. Try to go through one hour without making a judgment. Just try to make the coffee. The day is a good day because it is warm. Forget about the farmers who need rain. The day is good because you have the day off. The person who will need your special skill will not have a good day today. The traffic is bad. That means the economy is still good.

Try to motivate through the first few hours of your day without making judgments.

Do make observations. The traffic is light. Hum is the economy slowing or is this another weird State Holiday? My wife did not kiss me today. Did she have something to say but couldn't, but let me know that she did by not kissing me?

Not making judgment is not the same as not paying attention. In Truth, you will find that the brainpower that you have been using to make all those judgments can be used to pay more attention. It's like freeing up extra RAM.

— *Fields* —

Don't run the yellow light. Stop and Look and Listen. Maybe there is a butterfly to see. Maybe there is nothing. Maybe you are going to miss that speeding car that can't stop at the next intersection. Bless the Holy Field and all that you have in your presence. This is not easy. It is much easier to curse the darkness than to light a candle. The lit candle is definitely the way to go though.

After a while, a subtle thing will begin to happen. You will begin to see things differently. Your field will change. You will begin to see larger patterns. You will still have to deal with irritating ass holes of course, but they will take on a more archetypal feel. Small things will be part of big things. What you thought were big things will turn into small things that have no substance. You will begin to see how some small things are just the edges of really big things that are just beginning to emerge from the waters of collective reality. There is still no reason to judge. You are seeing only the field of words, the field of symbols, and the field of Man.

Observe. Listen. Feel.

See as much as you can. Hear as much as you can. Feel as much as you can. Discriminate, but do not judge. Practice not-judging with your lover, with your partner, and with the stranger on the street.

You will learn that you cannot find acorns in fields of strawberries. If you want acorns, you must find oak trees. Strawberries cannot wisely be judged for their lack of acorns. Persistence will not further in this case. You must find an oak tree if you want acorns. Hard work will not bear the fruit you seek. Waiting for the right time will not have its reward. Patience, in this case, is not a virtue.

Become like the Eagle. Keep your distance and keep your position lofty. Do not judge what is below. Know it and Watch it.

See the Same Way

You're looking at a picture.
I'm looking at it too.
Do you see what I see there?

Let's talk about the Difference
Find out what's in the way.
Open our eyes and see the same way.

I want to be there in that number.
I want to be there that Day
When we all,
When we all,

See the same Way.

Bruce Hornesby

SKY

– Sky –

The Sky is the limit. Or is it?

Or is the sky the beginning?

The sky is up.

But where is up? Up is down. Sky is out and away. Away from earth. The heavens are not up there.

They are everywhere.

Everyday, each of us is given a total view of the creation. True, we are only given half the view each day, but as the days goes by, we are given another piece of the universe that we could not see because of the blinding light of the sun. As the half year comes around, we see that part of the universe that was not visible just 6 months before.

It is so hard to remember that we are each riding on a spherical spaceship that is moving through space at great speed. As our spaceship orbits the sun, it follows the sun on its race through the galaxy. From the perspective of someone who lives on our moon, earth is constant and fixed. From the perspective of someone living on our sun, the earth is going in a straight line across the sky much like our moon. Of course, we are making circles around the sun. From the perspective of the Milky Way Galaxy, the earth is making a screw-like motion through the Galaxy. We are actually going through space like a coiled spring! From the perspective outside the Galaxy, both the earth and the sun are going in circles again. With the Galaxy moving away from the other galaxies in this expanding universe, the motion once again turns into a screw- like motion. From the perspective of the center of the universe, we are going in circles again. From the perspective of God, well you get the idea.

Whether in straight lines, curved lines, circles or apparently standing still, we are traveling at enormous speeds through something we call space. Our speed and our apparent movement is totally dependent on our point of reference. Our speed on the ball we call earth is breathtaking. At the equator, you are traveling about 1000 miles per hour, faster than the speed of sound!

At the North Pole, you are just turning in circles, spinning like an ice skater in a pirouette. That becomes clear to your mind when you see the sun go around you as if you have it on the end of a rope that you are swinging around you.

Then, there is the moon that is moving around the earth at over 2,000 miles per hour. Every 28 days it goes around the earth. Yet, is takes more than 29 days to go from full moon to full moon. Why? Because the earth has moved during that 28 day period to another position in space. The moon must go around the earth plus a little bit more in order to be aligned opposite the sun and the earth again. The earth is traveling around the sun at a speed of over 66 thousand miles per hour, over 88 times the speed of sound in our atmosphere.

With our feet on the ground, our heads are always tangent to the plane of the sphere we stand on. Our heads then are like antennae that stick out like little mushroom spores. With our feet and body grounded in gravity, our minds, operating at super-gigahertz speeds, are broadcasting light waves into the universe around us. Everyday our head is pointing out and seeing the whole creation. And the whole creation has the opportunity to see us everyday too. We are therefore connected at a physical level on a daily basis to the whole of creation and to the Holy Field we know as the Universe.

As warm blooded living creatures on this earth, we are taking the energy from the sun as found in our food in conjunction with the carbon atom, the energy in the hydrogen/oxygen atom combination and a mixture of various other atomic configurations and converting it all into light. (Or heat if you wish). And it's not just light, for what emanates from our brain is modulated light. It is Light with information. Hopefully, it is the Light of Intelligence. Thus, energy and the creation become conscious. We are carbon machines that accomplish a truly remarkable thing. From the so-called dust of the earth, we are able to make the universe conscious of itself, which might very well be the purpose of this whole affair.

And a lovely purpose it is indeed.

A very sophisticated observer from any point could or would be able to quote "know" our thoughts with reasonably unsophisticated listening equipment. The Universe is permeated by each and every one of our thoughts. It does take time, of course. But in a very real sense, the heavens know our every thought. And each thought is radiating into the universe endlessly. Every thought we ever thought is scientifically retrievable somewhere out there in the universe. Want to know what you said when you were four years old? Well, take your age, take four years off of it, multiply that by the speed of light of 186 miles per second times 8,766 times 3600, (the hours in a year and seconds in an hour) then determine what season and time of day it was and you can then predict where your thought wave is in the universe. If you had a listening device at that point, you could intercept that thought wave on its journey into eternity.

It is a little overwhelming to imagine that instead of being some insignificant piece of life on some insignificant rock in the vastness of space, you are actually a universal timeless presence. It's a little more than sobering. Your presence in the universe will forever be known and your effect on it will forever be felt. In case you ever wanted to be immortal, well here's the bad news. You are. It might make you wish you hadn't called that old friend who went small on you a worm. But in fact, in all fairness, he was acting like a worm. (Apologies to the worm phyla). There he is though, branded for all eternity, a worm, by you.

It's totally scary.

The realization that you are, in fact, a timeless universal presence does not come without its baggage. It's more like a container ship, a really big one. It's not easy to keep your focus on the important stuff when you are trying to beat a check to the bank. It's not easy when you are the Dali Lama, trying to get your kingdom back either.

Each of us is taking this assortment of atomic configurations found in carbon and water, combining that with the light energy of the sun. While we are in a field of gravity, and converting that into thoughtforms of light that radiate through all the universe for all eternity. It makes you wish you had acted better on the freeway today.

Not only are we timeless presences in the Holy Field, we are timeless presences in the present. We are Present in the Moment. Whether we like it or not, we are present. The real question is whether or not we choose to acknowledge that Presence.

The world of stuff behaves quite predictably and, often enough, consistently. Drop a rock in our gravitational field and it will fall. Drop a rock and a pebble and they will fall at the same speed. Go figure. Doesn't seem fair. Heavy things should fall faster. Who are you going to complain to? Set a rock on a table and it will stay there. It won't just decide to start rolling or jump off the table. At least it won't unless the earth starts jumping or you bump the table while looking for your glasses in the dark. Even though we know that the rock is just so much energy, we know that all that energy is going to behave and not go off on its own not paying attention to the laws that seem to define material reality. It is not going to fly off the table. It is not going to transmute into some other element. (at least not in local time)

Throw the rock and it's going to go forward and then down. It will hit the ground or the water and skip and bounce. It will come to rest. And it will not start up again just because it got bored or disinterested with the view. Most of what we are faced with in the Holy Field is pretty predictable. It's a lot better I think that living in the world or universe of "Q."

Sky

Light behaves pretty well. However, it doesn't seem to know what it is. Sometimes it acts like a particle and sometimes it behaves like a wave. According to some famous experiments by Michelson and Morely almost 100 years ago, it goes the same speed in all directions with no regard to the speed of the propagating source. That is pretty neat. Rocks don't act like that. If you take a rock and throw it out of the car, it knows it came out of a moving car. Throw a rock out of the back of a truck going 60 miles an hour at less than 60 miles an hour and it still goes forward in relationship to the ground. Light is a little more contankerous than that.

Light goes the same speed no matter how fast the truck is going. It's as if it's not paying attention to the local situation. Light goes 186 thousand miles per second whether the planet is going 66,000 miles an hour or not. It goes 186 thousand miles per second even with the sun moving through the edge of the spiral galaxy we call the Milky Way. It goes the same speed. At least we think it does.

Actually though, to be more accurate, it goes the same speed in a vacuum. Light going through air is slower. Light going through water is even slower than that. Light, at least the higher energy fractions of it, go even slower through things. Light going through thick glass is slower than light traveling through thin glass. Thus, you have your glasses. Lenses are interesting. They bend light and make all kinds of cool things possible. We can look deep into space with telescopes that are composed of lenses and mirrors that enlarge images. With the Hubble telescope and its corrected optics, we can see further and more clearly into the vastness of space and it's mysteries. With microscopes, we can look deep into the cells of our skin and see all kinds of things we would never have imagined.

Light bends when it goes through various substances because it actually decelerates. Materials then, act as filters and lenses that decelerate and thus bend particles of light. More dense materials slow light down even more. The highest energy particles can reach deep into the planet. One can intuit then that the highest energy particles are focused as they are decelerated into the center of the earth. Physicists employ this concept to isolate high energy particles coming in from Space. The unique geometry of the earth provides a lens for the highest energy particles of light to pass through. The earth then is a convex lens of gross material stuff that focuses the highest particles of subtle stuff. Conversely, the space between two celestial objects is a concave lens of subtle stuff that focuses gross stuff. The micro is the reflection of the macro.

In Santa Monica, California on the beach, there is a building that is right off Ocean Drive. There is a camera obscura inside. Now, the basis of any camera or lensing is a round hole. Don't ask me why, it just happens. Take a black box, put a small hole in one of the walls and the light that comes in will form an image on the back wall. I think it's marvelous. Anyway, there is a large camera obscura that looks out toward the water and the beach there in Santa Monica. But this camera invites you to come inside. It's exactly like being in a giant camera because you are in giant camera. There on the back wall is this incredibly rich view of the scene outside. The light level is less because it is all coming through a fairly small hole, but the detail is dazzling. It has just as good a resolution as it would if you were looking at it directly with your own eyes.

I guess you are.

Even more interesting, other negative shapes and orifices besides pin holes work too. Any kind of edge that encloses the light will act as an imaging agent. The leaves on the tree will do it. Look at the ground when the sun is out. In the light that comes through the leaves, you will see suns on the ground. It's like the creation is just dying to reveal itself to itself. The next time there is a solar eclipse and you want to see it, just look at the shadows created by the leaves of the trees. There, in between the leaves, you will see crescent suns by the hundreds.

The ability of material things and their edges to act as imaging agents for the rest of the imaged universe becomes even more confusing when we realize deeply that material stuff is actually not material at all, but a highly organized energy field. It is so well organized that at our level of observation and at our level of time, it appears as "solid."

There is a wonderful book called Factors of Ten. It starts with a couple lying together on the beach and then jumps by a factor of 10 into their skin and deep into the cells of their bodies. Then, it begins to take off and away into space like a rocket ship. The remarkable thing about the process is the repetition. Some times the view inside the cell looks like the view out in space. It's spooky how much empty space there is. It should make the real estate agent in you really nervous.

There is an enormous amount of space out there and in here. And it is full of energy. It is chalk full of high energy particles and wave forms and whatever else you want to call them. Think about it. It takes a lot of energy to hold your computer in place, but how much does it take to hold the sun in its place? We are talking about some serious lifting power here.

So you thought the sun was floating?

Floating on what?

Oh yes, I remember, it's a vacuum.

We all read that in school. Some part of me makes me think that I am more inclined to believe that Atlas is out there somewhere than I am inclined to believe some hocus pocus that requires me to accept that all this so-called heavy stuff is just floating out there on nothing. This nothing stuff must be pretty substantial stuff.

Sky.

Pie in the Sky.

But for now, forget about this and get centered. Get centered on the space ship.

The Practice

It really makes me nervous to think how many well educated people don't understand the earth, the moon, the sun, the planets and how we are traveling with them through this great adventure we call time and space.

The Spaceship

The best way I know to get centered on our spaceship is to make it a habit of watching the heavens and watching the sun go down, and then immediately throw that whole idea away. The sun is not going down. The earth is turning. The sun is no more going down than the earth is rushing by my window as I drive down the freeway. Watching the "sunset" is not quite so misleading, but I prefer something like "the turning into shadow." As the sun sets, we are turning away from it. So practice this on your first night. Watch the sun, but instead of seeing it go down, feel the earth turn. Feel yourself on a ball that is moving towards the east. Feel the earth turn. Feel the turning. Watch sunsets until you actually feel it. In my head, I add some strong deep *THX* sound track that accentuates the visual turning.

If you watch the earth turn at the turning into the shade of the earth's own shadow, take notice of the stars. Very likely you will see a large bright star slightly above the sun at sunset. If you do not, it is because Venus is in the morning sky. But first, you need to center on the moon. If you do not know when the new moon is, then look it up on a calendar. Several days after the new moon, you will see a crescent smile above the sun at sunset.

Each nightfall, watch the turning and what happens to the new and waxing moon. The moon is traveling around the earth every 28 days. Each night you will see the moon a little higher and a little more full of light. Don't confuse the turning with the 28 day cycle of the moon. In a week, the moon will be directly overhead at nightfall. It will be a half moon. That is the first quarter. Then, during the next week the moon will continue to be brighter and farther to the east as the sun slips from your view. In 14 days after it is new, you will see the moon quote "arising in the east" as the sun quote "sets." When the moon appears at sunset as the sun leaves your view, it is the full moon. The moon is on the opposite side of the earth from the sun.

The next night, the moon will appear about 50 minutes after the sun leaves your view. In a week, it will not appear until midnight. In another week it will not appear until right before the sun appears. It will again be just a sliver of light. As it moves into the sun, it will become another new moon. In a few days, you will be able to see it in the evening above the sun again.

To be centered on the local system, "feel the turning and understand the moon." Once you truly understand, you will be able to look out at the moon, see its phase, and confidently predict the next full moon within a day. You will be able to sit on a park bench and see the shadow of the bench move across the grass and know that your space ship is turning. You will be able feel the turning at any time, day or night.

The Solar System

Once you have found yourself on the ship you live on, you can begin to find yourself in the next context. And that context is the solar system.

We all received lessons on this when we were in second grade or maybe from PBS. While you watch the turning into shadow, notice the stars that are above the sun. Pick one or a group that you can identify and watch them grow closer to the sun as you watch the lunar cycle.

Much like the sun setting, these stars are not growing closer to the sun nor is the sun growing closer to them. We are moving around the sun, and the background is changing. It's like moving your head to see the movie screen when that big hairdo sits down in front of you after the previews and as the movie is just starting. The head in front is the sun and you are the earth. The screen is the background of stars. The earth is moving around the sun. We all know this, we just don't feel it, just like many of us don't feel the turning. Imagine when you see the sun that the earth is moving around it always to the right (in this hemisphere). In order for the earth to get around the sun in 365 days from an orbit of 93,000,000 miles, we must travel PI x D or 3.1414 times 186,000,000. (Yes, that number is the same number times 1000 that is the so called speed of light). That means we have to travel a little more that 584,040 million miles around the sun every year. Divide that by 365 x24 and you get 66,666 miles per hour. I don't know about Kilometers, but these numbers in the English system are pretty interesting. (I have chosen to forget the ¼ day).

So we are traveling around the sun about 85 times the speed of sound. That's about 18.5 miles per second or about one/10,000th the speed of light.

As we move through the year, watch the background move behind the sun as we make our way around it. You will begin to feel our orbit. If you are looking as we turn into the shadow, you may see Venus. Venus is the brightest star in the sky. If it is not just above the sun, then it is just below the sun, which means you can see it in the morning above the sun (which is below from the your viewing position in the evening).

Venus is between Earth and the Sun. It never gets very far away from it because of that. Mercury is between the earth and sun too, but it really never gets very far away and because it is so small, it is a rare find. But Venus is easy. Find Venus. Watch it. Notice with binoculars that it too has faces like the moon.

If you connect up Venus, the Moon, the Sun, and the Earth and make a plane out of it, you can project the plane of the solar system. Find the other planets and watch them. You will begin to see the plane of your own solar backyard. All of the planets travel in it. The Solar System is actually flat and you can see that if you try. Try a rooftop that faces south with a 3 and 12 pitch. With your head pointing roughly at the North Pole, your body is perpendicular to the solar dance floor. Imagine the turning and our trip around the sun. Don't just imagine. Feel it.

The Galaxy

To center yourself in the galaxy you must first find the Milky Way. It is a cloud of stars that runs north and south. Our solar system is on the edge of this galaxy. Our star system is rotating around a center that is located in the evening sky during the solstice in summer. Look straight up during the middle of the summer and to the solar dance floor. There you will see the center of our Galaxy. If you can feel the movement of our star system around that point, you pretty much know where you are.

You will never look at the stars and the moon and the sun again the same way. You will always know that you are a traveler through the cosmos on a spaceship that is moving through immense amounts of space and time.

And yet we know that all that matter and all that space is energy. It is some probabilistic display of electrons, fields, and photons, and something else called gravity.

It might seem odd to talk about gravity in the course of thinking about the Sky. But without gravity, there would be no sky. It would all be a big energetic mess. Perhaps there is no such thing as gravity. I used to think as a kid that there was no gravity, just expansion. Everything is expanding so fast that the smaller things apparently stick and are actually just appearing to be attracted to the bigger energy field. It is kind of like thinking that the sun goes down, and so on.

Gravity is a pretty big deal.

If this advanced technological scientific state we live in knew much about it, we would do things a lot differently. We could have little antigravity scooter-like devices like we used to see in Dick Tracey Cartoons for example. Einstein and a lot of really hard working smart people are trying to find a unified field theory that embraces all of what we know so we can understand gravity. According to Einstein, gravity is actually a curvature in space that occurs. And according to Einstein, Space is in a continuum with Time.

Then there is the idea of antimatter.

You can't have a dylithium crystal drive that gives you warp speed without understanding antimatter.

I find it better to not have such a need to name the Holy Field. In some ways naming it only confuses you more. You forget that it is a field of energy operating in time that is conscious or becoming conscious of itself. It seems better to remember that our eyes are seeing only a small slit of the radiation that is present. It seems better to remember that our ears hear only a small spectrum of the total sound that is present. It seems better to constantly remind yourself that you are not touching a table but an energy field present in that moment in time.

It seems better to stay in that moment with all of your senses and sensory equipment as turned on as they can be.

Sure, we need to allow our subject/object bicameral mind to name every thing or event our senses can perceive. But we cannot allow that naming to hide the splendor of it all. For, if we do, the Holy Field becomes a Godless moribund creation.

Then, instead of being connected to everything and knowing it directly, we become one who knows nothing about everything.

We will have forgotten that we live in a cave and in a world of shadows.

We will confuse what we have named with "what is." We will think that space is empty and that things are full. We will think that light resides on the things we see and on the retina of our eyes, but not in the spaces in between.

We will think that we see light when actually we see only what and where it has illuminated.

We will confuse shadow with darkness and light with day. We will flock to that light and not to the oneness. We, like moths, will find our attraction to light to be an invitation to folly and a circular yet sure flight to the loss of life force.

NAMING

– Naming –

What have I got Doc?

You have a case of synoptic aphasia.

Sounds horrible.

It's worse than it sounds. But we can treat it.

I knew I had something. Just didn't know what it was.

Well, now you do.

What do I do?

Nothing.

Nothing?

Nothing.

Don't watch TV. Don't read the paper. For sure don't read advertisements. Don't spend any money unless you must. Buy as little as possible. And don't talk to any friends or relatives who do.

I don't understand.

I know.

But at least you know what you've got. Just do what I say and come back after the moon goes into the sun.

When is that?

Figure it out.

How? Without the Weatherman? Without the Newspaper? Without the Almanac? How will I know?

Watch the Moon. Watch the Sun. Watch the Sky. This is part of your road to recovery.

What about medicine?

You need drugs... lots of drugs.

What Kind?

Almost anykind will do. Start with Pot and go up from there. Take Peyote. LSD is good for this kind of thing. Even good coffee will help.

You can't be serious.

I'm totally serious. Do you want to get well or not?

I want to quit thinking the way I do.

Then do what I say.

What else do I do?

Go on a strict "less legs the better diet."

What's that?

A fruit or vegetable or nut is better than fish, a fish is better than a chicken, and a chicken is better than a cow or pig.

What about Bread.

Man cannot live on Bread alone.

What else?

Run.

– Naming –

Run where?

Anywhere, in circles, on a treadmill, around the lake, to work, to the store. Run at least 10 K everyday.

Impossible?

Do you want to get well or not?

Remember, you won't be watching TV, you will have lots of time. And you won't be reading, or really talking to that many people. You will have time. It's only about an hour. Once you get the hang of it, you will look forward to it.

What's the purpose?

Remember you have an advancing case of Synoptic Aphasia. If not corrected now, if will be terminal.

But you said earlier to do nothing.

Nothing is good.

But this diet and this exercise is not nothing.

Your right, it's not nothing.

So, why did you say do nothing?

Because doing nothing will do. Doing more will do more.

You will understand. Go along now, and come back when the time is right. I have others who need to see me.

Do I need an appointment?

No, you just need to come back on the right day.

What time?

When you can.

Will you be in all day?

No.

Then how will I know when to come.

You will know.

I don't understand.

I know.

At least I know what I got.

Yeah, at least you know what you got.

Taking Names

Do you remember the first time you figured out that you had a name?

It would be really cool if you could. Imagine being part of a society that didn't name anything. Or maybe didn't name them until they were grown or of age, or some such thing. And I mean didn't name in the sense that you had no distinct identity until you got the name. You were simply part of the herd or family or whatever. You were not separate. You were part of the operating whole. Finally one day, the old men who did have names came up to you and said, "OK, come with us, this is the big day, today you become some one." You look around at your brother and sisters (or whatever they are) with more than just trepidation and give a few high fives and off you go.

Bango, You are now Something or more accurately Someone. You are now Ralph.

Congratulations Ralph.

– Naming –

You could tell the story about the day you were named. Yeah there was this big dark smoky room and I walked in and they said some magic words and suddenly I was who I am. Ralph. What a big day!

You are no longer some undifferentiated part of the community. You have a name. If you are somewhere and something goes wrong, someone will take names. Oops? You mean now I am responsible? Well kind of. You are more like blame able. Whether or not you are responsible is an entirely different question and issue.

I wish I could remember the first time someone said "Michael" and I looked up and said, "Oh, that's me. I need to respond." When I was 7 years old, I moved to another town. When I checked into the classroom, Ms Riley said, "Oh how nice, another Mike, you will be the 5th Mike in the class."

"They call me Micky," always one to think on my feet.

"OK Micky, take your seat."

I took me 5 years to kill Mickey. Even today, some still call me the endearing West Texas double name Micky Jim.

Names are powerful. If my name was Ralph, I don't know what I would do. I might want to be something really important. Most likely, I would change my name. I don't feel like being called something that sounds a lot like the noise that drunks make when their time of reckoning with their stomachs has arrived. Emerson did OK though with his totally ridiculous name. It drove him to poetry though.

If my name was Woodrow, I would probably want to get an advanced degree. My middle name is James. I always thought that was pretty normal. No reason to get upset about James. They are everywhere. Then, Michaels are everywhere too. There was that silly song about Michael rowing his boat ashore back in the sixties that made me nervous. Then there is the archangel

Michael. Makes it a little like being name Jehovah or Jesus. Maybe being named Michael made me want to save the World and fight the dragon and you know the story.

Names are important. Dogs respond to their names. Cats don't. Cats are transcendent that way. Many of us name our cars, and our computers. It gives them personality. Naming is high magic. Calling people names is not just impolite, its serious voodoo. Call someone a jerk or an idiot and they may never forget it or forgive you. Call your child a loser and watch the magic work. They either will or they will not be, but they will always remember the words. Call someone a name on the freeway and back it up with a gesture of some kind and you might get shot. (at least here in Texas)

We take names for granted. Like most everything else that becomes part of our day to day consciousness, names are taken lightly. Now, most of us are just numbers. In a way, it's better to be a number. It kind of takes you back to the time when you didn't have a name. You just have an ID number like a new car or camera. But we do have names, and they make us more than we imagine. We used to debate that is wrong to name a child until you saw them. How could you name someone without seeing them and holding them, and looking deep into the portal of their soul? Some of us name our children with naming books published from Hallmark or something like that. Or we get named from an Uncle or a friend, or a folk rock singer if you are really unlucky. There were a lot Chelseas around thanks to Joni Mitchell I think.

Maybe all of the Michaels came after the WW II when every one saw the destructiveness of the Atomic Bomb and we just hoped that an Archangel would come save us from ourselves.

Names are important. They are big time Voodoo. It used to be a really big deal to name some one or thing. It formed them. They formed into it. I have never known a Rupert that I liked. Seriously, most Jerry's are goofy and most Leroy's are either singers or welders.

– Naming –

Georges are generally pompous. Too many kings named George I guess.

However, the Louis(s) are not such a bad bunch. My brother is a Frank. And he is not. Most Franks act like the French or the Frank on Mash. To be Frank is because "they" are not. It's a mystery.

Susans generally look like Susans. I don't why that is. It is the same for Marys. In fact, a good way to remember people's names is to actually listen to their name when they tell it to you and then think of all the other people you know with the same name. It's remarkable and a little weird to see the similarity in these named groups.

I mean, come on, how many Bruces do you know that are gay. And Robert? Love that Bob! Of course there are exceptions, the fact that there is a rule is the point I'm trying to explore. Names are important.

What do I have Doc?

You have synoptic aphasia.

Most of us can't even get well if we don't know what we have. Ask any doctor. More than half the job of healing is naming the disease. Most of us can't accept an event unless we can name it. The space shuttle blows up and we have to know why and we must put a name to it that explains why it happened. It was the left wing. It was the foam insulation. It was a sign from God.

Naming is an essential part of our bicameral subject object mind. It is the basis of the destruction of the Holy Field into a bunch of unholy alliances.

Michael sits in the chair on the floor in the house on the land in the city of Austin on the planet Earth held together by a gravity that we don't understand and warmed by a light that acts like a particle but sometimes behaves more like a wave. I write with my fingers on an interface called a keyboard using a system of names. You read with a subject/object mind and take in these symbols. You get meaning from these symbols.

But the poet plays with these symbols. He uses the names to actually say something else. As I often say, don't listen to my words, listen to what I am saying. Shakespeare did so much more than just write words. He wrote words that carried the weight of reality and the magic of the Field we find ourselves moving through.

We confuse knowing a name with knowing.

The Naming Tool

Naming is a good tool. Like any other tool. you can become lazy if you use it too much, or you use without understanding what is would be like if you did not have the tool. Try digging a hole without a shovel. It will make a believer out of you. The Knife, the Wedge, the Lever, are all cleverly combined to make a very effective device for moving dirt and other such material. Try opening a can without a can opener. Try using your hands alone, or try using a rock or a blanket.

And there are other ways to make holes too. You can use a stick of dynamite. You can use a back-hoe. These tools can be remarkably superior to the shovel.

There are other ways to understand things without naming them.

Yet we constantly fall prey to our own devices.

Imagine going through your day without naming everything. It is totally foreign to your intelligence. It is your intelligence.

The Practice

Try to get through an hour without naming. When you see your sister, see her for what she is. Look at her. Look at her as if you had never seen her before and you know nothing about her.

Look at your car. Don't see it as a Chevy. See it without naming it.

See the color. See the dents. See the tires that are worn. But don't name it.

If you get through an hour, you can get through a morning. If you can get through that, you are well on your way to being cured.

Without the naming, you then can ask?

Who is the you that is thinking?

Who is the one who would watch you not name?

From the beginning of your Field, you have been programmed.

Your operating system, so to speak, is one of names, of individuated reality. You are Frank, or Bob, or Bruce. You have brothers and sisters and mothers and fathers and friends and enemies.

The great Sage, Ramana Maharshi, constantly questioned this.

Who are you?

Are you your name?

Are you your Body?

Are you your brain? Or are you the thoughts of your brain? And what are these thoughts? Are they not just ripples of modulation on the light of consciousness?

From the perspective of our religious teachings, are you your soul?

From this perspective, who goes to Heaven? Who goes to Hell?

What part of you is immortal?

Certainly, the part that is immortal is not your body. It is not your brain. What it is then?

Who is reading this? Who is watching you read this?

WHO?

What is the Name?

Our operating system just can't compute without them. It's like trying to get Windows to work without using Dos. The whole system is based on it.

We just don't know how to make sense of it all without them.

Yet we know that the early descendents of Abraham were cautioned to never speak the Name of God. It was too Holy to speak.

When we name the Holy Field, we maim the Holy Field.

What is gained in the middle is lost on the edges. As we describe our reality in terms of names and events and time, the magic of it all disappears. Instead of flying through the Holy Field like eagles, we creep through it like a serpent. True, the serpent has a visceral feel for the earth and the grass and the

– Naming –

things of this world, but it does not have the view of the Eagle that flies over the world seeing it wholly and as one.

The oneness if lost in the countless naming.

Naming allows us to break the Holy Field into infinite parts. By naming we can manipulate. By naming we can find cures to the diseases that bring us to our own mortal ends. By naming we can build spacecraft that can cross the solar system. By naming we build our homes and our lives and our future. By naming we can have power over the Holy Field.

It is a high sorcery indeed.

Consider the consciousness of a Bird, or a Lilly.

They name not and they toil not. They simply are.

WE can name light and predict its speed as it moves through the Field.

We can even predict that as you approach the speed of this light-energy, that time itself will be changed.

We can talk about gravity as a force. We can pretend that we understand much of what we see and know, but in fact, we know nothing. We know as much about that that is outside of our sense of self as we know about that that is in fact our self.

The Teaching

According to our religious teachings, "the Lord God took the adam, and put him into the garden of Eden to dress and to keep it. And the Lord God commanded the man, saying, Of every tree of the garden thou mayest freely eat; But of the tree of the knowledge of good and evil, thou shalt not eat of it; for in the day that thou eatest thereof thou shalt surely die. (spiritual death)

Thereafter in the next five verses, woman is made from the rib of Adam.

Marriage is instituted in the next two verses.

Then, in verse 25 of the second chapter of Genesis, the Bible states, "And they were both naked, the man and his wife, and were not ashamed."

Next, in Chapter Three, comes the drama, "Now the serpent (Satan) was more subtle than any beast of the field which the Lord God had made. And he said unto the woman, Yea, hath God said, Ye shall not eat of every tree of the garden?"

The woman replied, We may eat of the fruit of the trees of the garden; But of the fruit of the tree which is in the midst of the garden, God hath said, Ye shall not eat of it, neither shall ye touch it, lest ye die.

And the serpent said unto the woman, Ye shall not surely die; For God doth know that in the day ye shall eat thereof, then your eyes shall be opened, and ye shall be as gods, knowing Good and Evil.

And when the woman saw that that the tree was good for food, and that it was pleasant to the eyes, and a tree to be desired to make one wise, she took of the fruit thereof, and did eat, and gave also unto her husband with her; and he did eat."

The rest of the story is, so to speak, History.

And we have become, just like Satan said, like gods.

And we have, just as God promised, surely suffered a spiritual death.

But there is more than just the tree of Knowledge, there is also the tree of Life. Partaking of it was of particular concern to the Lord God.

"And the Lord God said, Behold, the man is become as one of us to know good and evil; and now lest he put forth his hand, and take also of the tree of life, at eat, and live forever; Therefore, the Lord God sent him forth from the garden of Eden to till the ground from where he was taken.

So he drove out the man; and he placed at the east of the garden of Eden cherubim, and a flaming sword which turned every away, to keep the way of the tree of life."

In the next verse, Adam *knew* Eve, and eight verses later, we have the first homicide.

Imagine how powerful the first thinkers were. They knew what was happening. They knew how things worked. They could repeat things. They could direct the efforts of others. They knew how to start fires and find the food and water.

Very heavy thinking was left to the priest and those who would hold the keys to the tribe. After a while, it became necessary for the priest to scratch some of what they knew down on a rock. They had to invent drawings that would capture their thinkings. Reality slowly and surely became replaced by an abstraction of reality that existed in the archetypes and names in our system of understanding that ultimately is now known to be knowledge.

In this knowledge of names and archetypes, we have lost touch with the original foundations of the Holy Field. We have been cast out of the garden because we were tempted to be as gods. Our Consciousness of Good and Evil has made us into gods that are spiritually dead.

How do we make our way back to the garden?

Or do we even care?

Now that we have come this far from the fall, can spring be that far away?

Our present consciousness is like the operating system that runs the computer I am using to make the marks that record these thoughts. It is built upon code upon code upon code. This operating system based on naming and good and evil and everything else that comes with it is like the operating system that has become too top heavy and way too likely to crash.

Perhaps it is this eschatology that we face.

The end of the world it not the end of the world.

It is the end of this world.

This world of consciousness that has served us well enough over the millennia but now is simply overburdened with so much code that we simply cannot see the proverbial forest for the tree is headed for the trash heap of history. A new operating system is being developed by the planet.

Naming is a Tool. This naming consciousness that we all share is a tool.

It is not Reality.

Reality is the Holy Field.

Subject\object consciousness is an abstraction of that Field. It is not the Field, just as a picture of a great tree is not the tree.

The picture will not cool you from the sun, nor will it drop its fruit to the ground. The abstraction is not the Reality.

We are beings of energy participating in a mysterious Field of Energy and Time. What we have here is a genuine class nine miracle! It is well beyond our power to add or detract.

To know that deeply and profoundly connects us to the power and presence of that mystery.

– Naming –

To name it countlessly and incessantly leaves us fragmented and unrealized. In the words of the Teaching, we shall surely die.

We have synoptic aphasia.

SEEING

– Seeing –

hat do we really see?

Do we see Light?

Actually we do not. Try a simple experiment. Put a flashlight on one end of a shoe box and let the light pass through to the open side. If there is no dirt floating around in the air, you will see nothing. Light is quote, "invisible."

Look into the dark night sky. Between those stars and planets is darkness. Yet you know, at least in your mind, that it is totally full of light. When the light has nothing to hit or illumine, it stays invisible.

Now, you can argue with me and say, "Watch what happens when I stick the flashlight in your eyes." I would say that I am seeing the reflections of the mirrored surface and the energized filament of the light bulb itself.

This is not an angels on the head of a pin issue.

What are we seeing when we see?

We would not see anything if it were not for materials which are, in fact, just bundled up energy. So, in order for us to see the creation, then there must be a creation. This is more than just tautological. It takes energy striking bundled up energy in order for the creation to reveal itself. And it takes someone or some other living thing with eyes to see it or it doesn't make a lot of difference anyway. It's sort of the tree in the forest noise thing.

From a physical point of view, we actually see very little. We see with our eyes a very narrow strip of the electromagnetic spectrum called visible light. A mosquito would argue with that definition because they can see infrared light too. Consequently, they can find you in your bed at night even with the lights out. So can our soldiers with infrared goggles. To a mosquito or a soldier with such goggles, there is no such thing as night.

Warm things are everywhere. Various creatures of the creation can see different parts of the electromagnetic spectrum. Cats see well in the dark. So do owls.

Bats see with sound waves. The concept is slightly different because they emit the sound. It's like having your headlights on a very very dark night.

If you look at clouds carefully, you will see that part of the cloud is moving but another part of it is just disappearing while another part is appearing. The cloud is actually a way to track what you are really looking at, which is an energy form. The clouds are what happens when the energy form moves through the atmosphere.

In a like manner, objects and life forms are energy forms moving through the Holy Field. It is relatively easy to see that a cloud moving through the sky is really just an energy form moving through the atmosphere, it is quite another to see your sister moving through space\time as a living energy field moving through the Holy Field. Yet, we know that the body is full of space just like the space between the galaxies.

This seems totally ridiculous when you are avoiding the energy field in the form of the SUV that is backing into you in the parking lot. Energy fields act a whole lot like objects and they don't just revert back to the way they were. In fact, its kind of important that it stays the way it is bent. Otherwise, we would be in a heck of a lot of trouble. If everything had memory of its former state, we would have some real problems trying to make anything out of anything.

But I digress.

The point is this, we see a very narrow slit of reality. That very narrow slit is then interpreted through a very specific subject object matrix that takes the light show and turns it into an understandable reality. However, the reality is just an abstraction of the reality. If we saw and could understand like the Shaman, we might see something entirely different.

Seeing is uniquely tied to time. Time lapse photography reveal plants that wave and almost talk as they grow. Laser illuminated photographs of rain drops reveals a unique world that our eyes of normal time cannot pick up. The advent of the moving picture demonstrates this relationship. Flicker images in front of the eye and they take on life. The eyes and the brain blend them together seamlessly, even though we know that the pictures have no life of their own. Only when they are presented in time do they have life.

Time lapse photography shows buildings that grow, canyons that are formed, cities that breathe with whizzing lights and a blur of people walking through a blur of streets.

Seeing is truly a mysterious thing.

Shamans speak of the first attention. It is the attention that dominates the mind of the world. The first attention is hypnotic. There is also a second attention. This attention breaks the hypnotic spell of the first attention.

The second attention is a different way of seeing and of being. To access this second attention, you must be willing to let go of the first. You can do it anywhere. But it will most likely reveal itself briefly. It may come with joy, or with fear, or with music, or in a movie. It is the oneness. It is the oneness that we all seek and crave. We somehow know that we are born of it and that we have grown out of it.

The Practice

There are many ways to find the second attention.

Here are a few from Dr. Jose Stevens:

Try staring at your reflection in a mirror and make direct eye contact for twenty minutes or longer.

Take a walk and focus on the shadows only.

Or walk with a very soft focus and use all of your senses to absorb everything around you. Try to bring everything to a whole. Bring the Field to a oneness.

Or focus only on the background of the objects in the field of vision. Instead of seeing only the branches of tree, focus more on the spaces between the branches of the tree. Focus on the space between the clouds, the buildings, the people in the restaurant.

Or perhaps most interestingly, using your imagination, see the woman in every man and the man in every woman you interact with. Once again, you find the oneness in each of us.

You can practice being in the second attention while using your first attention. Then you can walk between the worlds. Then you will see each moment, each thing, each event in the Holy Field as a miracle. For it is.

Once you can see the Oneness, then you become a conscious part of the Oneness.

Then you see the connectedness.

No longer is the tarot some mysterious deck of weird pictures that somehow foretells the future and explains the past, it is a working piece of the oneness. It would be odd, if it did not work.

— *Seeing* —

Seeing is the faculty that allows us to be one with the creation in the primary language of the creation itself. When we look into the night sky and see stars that are literally hundred of light years away, we are seeing the creation of the past. Some of those stars may be gone in the time of this earth. Some of them might have exploded sending a lethal death ray of radiation that is hurling upon us and we would know nothing about it.

Some other species from another galaxy might know about it though.

Seeing allows you to see not only deep into the creation, but it also allows you to see deep into your fellow passengers on this starship.

Not too long ago, while writing this book, I was sitting in a particular coffee shop in Texas. It is a good one. It seems like everyone who comes in through the door is beautiful.

One morning, I sat waiting for Martha, my South American partner. (at the time)

Suddenly, as though a light was turned on, I could see the emotions of each person. There, in their face, each one projected an emotional field from their face and eyes! Obviously, they each wanted coffee or at least a bite of the wonderful pastries that are made there. But I could see more.

There was the woman who was worried about her life. There was the man who worried about his job. There was the happy child. Then, in walked Martha. She was proud. She had solved the problem of finding a gift for a young child and difficult to please parents.

For almost an hour, I watched the emotions of the people as they came and got in line to get their coffee. Occasional tears fell down my face. I was becoming attuned to the emotional fields of everyone in the room. And these are very unusual free thinking spirits to say the least.. I thought about the Gurus who would see face after face and know more about the people behind those faces than they knew about themselves. What seemed a mystery at the time was now no mystery at all. It's obvious. People wear their emotions on their faces. I must add that this place is not a coffee shop in Dallas where fake smiles are pasted on faces like the fake red lips and the fake eyelashes. This coffee shop is no carnival of weird creatures. These people are real. They are not fakes.

After about an hour of this, I began to imagine a Twilight Zone story.

A man is mysteriously endowed with the ability to read people's emotions but ultimately finds that it is more of a curse than a blessing. You know how the story goes.

For me, it wore off like a beautiful perfume. The scent still lingers from time to time, and if I need to, I know where the bottle is.

It makes me a better human, a better lover, a better writer, and a better better when we play the "who is why is" game in the restaurant. I think most of us can do the same.

We can all see our loved ones when they are happy.

We can tell when our boss or our co-worker is bothered.

It's no big deal to know or sense when something is not right.

The truth is, "to not see" is what is abnormal.

And that it is the condition of the First Attention.

The Second Attention reveals the depths of the Field to anyone who would reside in its nature.

More Practice

As you begin your day, imagine that you are a Shaman who has come down from the mountain and that you will spend this day with those of the First Attention. You will treat everyone with respect and dignity.

See how their faces move.

Then see how they walk. Look at their desk or their shoes. See into their eyes without making them nervous. Look at their hair, and at that make-up. Watch for twitching and other subtle movements in the muscles of their face. Of course do not judge.

Observe.

If you observe with a judging mind, you are only observing yourself.

Watch how they become nervous and behave in front of the boss or a new potential boyfriend. Watch how they act around their children. Watch how they meet new friends and notice if they remember the name of the person they just met a minute later.

They reveal everything about themselves.

They are the proverbial open book.

If they smoke, watch how they smoke and when they smoke.

Do they tap the table with their fingers?

Do they move their legs under the table like Thumper the Cottontail?

Are their arms crossed when they talk to the new hire and do they go to one leg when they address their co-worker?

Observe the way your best friend meets you.

Observe the way your enemy greets you.

Make a note of the first things your enemy says to you. It will not be a lie, no matter how hard he tries to tell you one.

Without being too self-conscious, observe yourself. But mostly observe your breathing and make sure that you are breathing from deep in your abdomen. Keep a soft focus, yet at the same time notice the detail.

Shamans and other sensitives of the second attention not only see all these things, they also see the light that surrounds each form. Everything that is above absolute zero is radiating energy. Remember, if a mosquito and an American soldier can see you in the dark, maybe a shaman can see you too.

People who know the second attention can see these emanations. Science has learned that they can even be recorded.

Learn to look at the edges of things.

See all you can.

See the sky and the birds. See the Sea

See the mystery and the see the beauty.

See the Holy Field and the oneness.

But mostly, stop seeing yourself. Go look into the mirror and stare and stare and stare until you have seen yourself enough.

Then go outside and see with eyes that see. See your wife with new eyes. See your children and your neighbor with a newness of spirit.

See the wind by following a gust of wind as it moves over a meadow of grass. See the wind as it moves through the trees to the trees across the street. See the clouds as they grow and die through the air.

See your self as a Shaman.

For you are one, if you choose to be one.

Seeing the Light

It's a common phrase.

When you finally realize something, you "Saw the Light."

"I see the light" is a common expression in our culture. When someone has a religious epiphany, they have seen the light. When you suddenly see the solution to a problem, you see the light. When all is not lost, there is a glimmer of hope. There is the light at the end of the tunnel. (hopefully not a train) When someone tells about their transition from the First Sight to the Second, they often use words like, I saw the light.

Others speak of their mind being filled with light.

But, like I said at the beginning of this chapter, you can't see light.

How can you see the light if you can't see light?

Today it occurred to me, that maybe we can see light.

It occurred to me that we all see the light, even if our brains tell us that we can't.

I have always been impressed with stories of the power of the brain in conjunction with sight. You know what I'm talking about. There are those studies where an upside down image rights itself in the brain. On a more accessible level, who of us has not put on a new pair of glasses thinking "oh my god, I will never get used to these" and boom, in a matter of hours, you have forgotten all about the weird, out of whack, visual world that existed just hours before. And bifocals, who could get used to them? Just about everybody. The brain is amazing at taking the information from our eyeballs and making it useful.

Maybe in the beginning, all the brain saw was light. And the brain said, "this is not meaningful." So it decided to not see light anymore, just where it has been.

That is why, when people become illuminated, their brains are filled with light. I think all of our brains are full of light.

We don't see light because our brains filter it out, We have become one with it.

We are light. We are light beings.

To see the light then, we must see ourselves.

We must be ourselves. Truly.

Then we can

"See the Light."

The Teaching

And God said, "Let there be light."

And there was light. And God saw the Light, that it was good.

And so are we.

This is not judgment. This is our nature.

BEING

We are human beings.

We don't say it that much. But that is what we are.

Beings.

And it seems that we are not very good at that.

We are better at being a thing.

We are our name. We are our family. We are our personality. We are our experience. We are our education.

We are our parents. We are American. We are New Yorkers, or Southerners. We are our profession. I am a doctor. I am a teacher, a writer, an athlete. We are what we are.

But we are really beings. We are often good, sometimes bad at being.

I am being a writer at this minute. Later, I will take a hike through the mountains. I will be something else then. At my best, I will be a Shaman who walks through the wonder. At my worst, I may worry about my unleased apartment 600 hundred miles away in Texas.

I can be very good. I can be very bad.

Bucky Fuller used to talk about being a verb. "I seem to be a verb," he would say. He knew we all were nouns.

He knew we lived in a prison of our own making. We fashion this prison for our own protection and for our own ease. It seems that most of us choose to stay in our mind prison- not because we seek protection from the other, but unfortunately, because we are a little lazy. It takes so much energy to be truly alive and in the moment.

It takes so much energy to constantly allow yourself to be changed as you pass through the Holy Field, yet when we do, we gain more energy. It is very much like Love. The more we love, the more love we have. The more we stay alive, the more life we have. The more we stay open to the Field and allow it to reveal itself to ourselves, the more we become one with the oneness and the more likely we will be the being that we are.

But it takes energy. It takes constant attention.

To be or not to be. That is not the question. That is the prison. To become or not to become is the question.

There is some powerful force of nature that operates against the ordered chaos of the Field. Perhaps we can just call it gravity, or ordering. Perhaps it is simply fear. But this is not the same process that is spoken of in Western Science. Here, there is a tendency to go from an ordered system to disorder, from order to entropy.

The world of the Shaman is more one of ordered chaos which moves to lifeless order. The fight is against death, less freedom, less openness to the oneness and the Holy Field. The Field is not chaos. It is maximum order under the mask of disorder.

But how are we to be?

We seem to know how to act. Most of us learn in kindergarden. Don't be selfish. Don't hit. Don't make people cry. Don't tattle. Share your toys. Be Nice.

In the grown up world, this translates into a person who has a quiet assured self-confidence. A well-behaved grown-up respects others. He has a good-natured sense of humor. He is unpretentious.

According to Jose Stevens, a well-developed being is more than this. If he has power, real invisible power, he will somehow display a commanding presence. He will have an ease and comfort with physical contact and eye contact and he is comfortable in any setting. He has natural leadership qualities that allow cordial and comfortable ends to conversations and meetings.

Even though he speaks with authority, he speaks with compassion. He is skillful with people and problems. Like Socrates, he asks questions that lead to discovery. He treats people equally without regard for position or status. He has a direct and compassionate honesty. He speaks clearly and simply in words that everyone can understand.

He smiles with his eyes. His dress is most likely simple. He knows how to clear away physical obstacles and apparent barriers in order to deal with others directly.

Who of us would not be this being?

But the trick of course is this:

If you try to act like this, you act like something else. To try to act like a compassionate person when you are not is to be totally fake. It would be better to be uninterested! To try to affect these mannerisms would be totally unauthentic and unappealing at best.

So to be real, you must be real. And the "real you" must be empowered by you to be this being. There are little rules of course. Four such rules come from the Shaman Michael Ruiz whose Four Agreements has received large acceptance in the popular culture.

His rules tell you to "Be impeccable with your word"…don't lie. "Don't take anything personally"… don't wear your feelings on your sleeve. "Don't make assumptions"….you will anyway, and "Always do your best"… something we should all try to do.

Jose Stevens tells us to, "Keep our perspective, stay flexible and go within" often. He advises to "pay attention to the details of life, intend to be highly aware, be willing to change, and be constantly aware and observant of small, apparently insignificant acts."

He warns to avoid the traps of responsibility, dependency, anxiety, distraction, self-deception, negative thinking, loss of vision, loss of detachment, and going on automatic which is the road of laziness.

But these are just words.

How do you feel them. How do we make these words and concepts part of our Being?

Waiting

I hate waiting.

And yet sometimes it seems that that is all that I do. I hate waiting in line or at a light or for my turn.

I remember when I was young. I could not wait for Christmas. I could not wait for my friend to come spend the night. I could not wait for the weekend and my date with my new love of the season.

Even today, I wait.

When I walk this afternoon, I will find myself waiting to finish the walk. Sure, I will be very in touch with the birds, the rocks, and the sky, and the mountains around me. I will speak with them and they will speak with me. Yet I will drop out of the consciousness and become impatient. Instead of losing myself in the oneness, I will think about the long hike back, and "will it get windy", and "did I bring enough water, or perhaps too much."

I will find myself very in touch with nature and the invisible forces around me. Yet I will fall away like a stone on a ledge.

I will say to myself, "There you are doing it again."

There you are not here, but there, that there that never comes and that there that never will and that there that is not real except in my distraction.

And then I found myself.

You must want to be aware.

I have friends who are coming to stay with me in the house. They are not here yet, so I wait. I will not be more happy when they arrive. It will be an event. The event will be what it will be. They may be happy, tired, excited, hungry, disappointed, or on some other unimaginable course.

Shall I attach my prosperity of mind to their arrival?

Who is the host here?

But tonight, I will think that I will be more happy when they arrive, because I am without constant attention from friends or lovers right now. When they arrive, I will be happy.

So I wait.

And, I am wrong.

And so I correct my error thinking. I go sit in the splendid sunshine. I write more. I feel the earth turning. I imagine my new book and how the contents will work. I notice the smallest of things. I hear the small lizard as he moves about in the trash. He seems to live there now. Then there is the bell, the beautiful bells that caress this place. The wind picks up as if to sing with them. The lizard dances.

And I go a walking.

Walking

Today I walked up to the Pueblo Fantasma. From this Pueblo, it is due east. The old road leading up the steep cliff is actually the original road that led into this silver mining town of two centuries past. The road is indescribably steep, yet most of the rocks in this cobblestone creation are more or less still in place. The builders used a unique system of rock placement with long stones making squares on about 4 foot centers and diagonal stones from the corners. In between, are the rounded cobblestones. Some of them have been there for more than 200 years.

I go up about half way and then take a trail that I had looked at before but had never tried. It leads me to the ghost town on this 10,000 foot valley between the Mountain of Talking Sticks and the Montana Sagrada for Gueros. These are my names. They must have others.

Soon, I find myself headed up the valley where the earth turns sidewise.

Instead of continuing that route, I decide to follow a little goat herder trail to the right. It takes me around the mountain and into a canyon and back around to the opposite side of the canyon I came up in. The mountains here are steep. One false step and you will find yourself looking for the brakes. The cactus, the wildflowers, the poppies, the agave, the giant sticks of asparagus that grow out of them cling to the sides of these slopes. Looking down at the ground, I see reds, and blues, and purples. These are not dull hues, they are bright and alive.

The rocks in this part of the canyon are purple. In other places they are bluegrey, and yellowgold, and then there are the jetblack ones with white lines in them. There are white crystals shining in the sun. They are dark rocks with little glints of silver still left in them. It's as if some great giant graffiti gang came through and sloshed color around hoping to confuse the people and the scientists that would follow.

As I follow my trail I am encouraged by manure. Animals use this trail. I am an animal. This can work. Sometimes the trail blends in to the rough evergreen ground cover that even the goats can't eat. Yet it comes back. Every now and then I lose it. Then it reappears.

My high tech walking stick is a must. The shale in the rock ledges breaks off easily making an extra leg a real important asset. Often it has saved me from turning an ankle and making things harder than you want them to be.

As I come around the last mountain side, I see the pueblo with the giant church that silver made. The dome is magnificent in the western light.

I am coming into town on a trail with a view I have never seen before. I pull out my cardboard camera and take a few shots. They will be dim reminders of the glory of the view.

I make my way down into the crowded parking lots. I must get by a donkey who has been parked in the trail coming down from the mountain. There must be at least a hundred working donkeys in town. They are like pickup trucks in rural Texas. Sometimes as I come down from the Holy Mountain of the Huicholes just a few miles away, I run into a donkey loaded down with flowers picked from the high meadows that lay in between the great rounded forms that make up the Sierra Catorce Range.

Real de Catorce is like few other places. Pilgrims come here by the thousands to pray and seek relief from their sufferings. Many crawl on their knees. They come to pray to St Francis.

The narrow main street is a carnival of Mexican religious kitch. There are thousands of plastic framed pictures of saints and angels and Jesus.

— *Beyond* —

There are clocks neatly fitted into Da Vinci's last supper. Some of the world's finest religious art is illuminated by a single Christmas bulb and framed in a golden orange hued plastic that is, in itself, a wonder. There are coffee cups with all kinds of writings. There is candy…lots of candy. One girl was selling her Dulce de tres leches. Women bark out that they have food for sale from their hot grills right on the street. They have gorditas, caldo, and much more. The street feels more like something you would find in a bazaar in the Middle-East.

Some vendors are selling rocks. Their signs say that these rocks can heal you. There is jewelry and key chains and CDs and T shirts. The blue plastic tarps lend their special hue to the street that is virtually covered from one side to the other as the puestas turn into a river of color and smells, and sounds.

Middle aged daughters walk with their grandmothers. Old men are wheeled in by whatever contraption is available. Families of five generations walk the polished almost shiny black cobblestone streets to see the Church that their great great grandmother knew when she was a young woman.

As you get closer to the huge church, you see more and more candles and milagros for sale. Milagros are the little metal arms and heads, cats and cars that are used for charms. Then you see little stands with green herbs in brown boxes. The signs say these herbs can heal you.

The church, though magnificent, needs repair. It reminds you of the great churches of Europe. In the back, there are hundreds of little metal sheets tacked to the walls with stories drawn and written about some healing event that occurred somehow in some relation to the Saint of this Church. Saint Frances sits in a glass altar to the left. The people crawl to his worn serenely surreal plaster form. The altar is illuminated by purple neon. Even though the sign asks that you to not light the candles, there are 30 or 40 going next to the altar. The pilgrims light them as fast as the church alders can put them out.

The church bells clang out the time of the next mass. I can see the young boy as he is positions himself in the bell tower from my window that looks to the East. I see the movement of his arms as they pull on the ringers of the two giant bells. A half second later, I hear the clang. He has quite a repertoire. And he has been busy this week with all of the masses. It is Christmas.

I am not waiting now.

The Teaching

How are we to become what we are in the oneness?

How do we stop ourselves from becoming these selves that are primarily an invention of fear and laziness. (and experience) And perhaps more importantly, can we come back from the grave, so to speak, and become alive once again, knowing we have allowed ourselves to become just shadows of our true nature.

We can.

We can change. But we must want to change more than we want to be the self we have created.

Quoting from Robert Pollack's book, *The Missing Moment*, "About a century ago, a close examination of the ambiguities of time in the content of dreams and daydreams led the Viennese physician and experimental psychologist Sigmund Freud to the clinical methodology he called psychoanalysis. From his clinical observations, he came up with a series of models of the mind that included-for the first time-an unconscious component to all mental functions, including the rational ones that seemed least likely to have any relation to the unremembered past. The strategies of psychoanalysis subsequently devised through trial and error by Freud and his followers have acquired a certain mystical patina. Actually, psychoanalysis is rather straightforward.

"It is based on the clinical observations that talking and listening carefully to a person's unguarded ramblings can help him to safely and reproducibly bring painful and embarrassing memories out of repression into consciousness; as careful and reflective conversation bring these memories to consciousness it also uncovers the hidden emotional connections between current and past difficulties. The purpose of this exercise in memory recall is also straightforward in clinical terms: to help a person to learn how to release the past's control of present emotions, actions, and beliefs. Once the underlying emotional connection of the past with the present is understood, the emotional content of the current difficulty-now understood in terms of earlier events- can in many cases be brought under conscious control.

"The notion that a person's destructive, self-defeating behaviors and disturbing dreams may be conscious manifestations of otherwise repressed and unconsciousness impulses gave childhood itself an altogether new somewhat ominous aspect. Many turned away from psychoanalysis deeply offended, and some still do."

In this model of conscious and unconscious thought, an inner trinity of contesting, unconscious mental states was developed. Here the id, the superego, and the ego were named. The unconscious ego is the part of the psyche whose conscious manifestation is a grown persons sense of himself. The unconscious id is the pool of motivation, and the superego is the memory of idealized authority, setting the standard of allowable thought and behavior.

If we are to become part of the oneness, then the superego must be modified to set a new standard of allowable thought and behavior. The subject object manifold of reality must metamorphose into a new model of reality. Our subconscious id must be full of the motivation to be truly alive. And our ego must be redefined in light of the realization that the Field and the Self are one.

– Being –

Krishnamurti, the Westernized Indian Sage, would say, "The Observer is the Observed." His Flight of the Eagle is no different than the flight of the eagle of the Shaman.

The Practice

It is important to constantly remind yourself that although the Field is conscious of itself, the manifestations within it may or may not be. Part of the mystery and beauty of the manifestation is the revelation of the unconscious into consciousness.

And so it is with each of us.

Most of us do not have the money or the time to invest in the psychoanalysis of Freud. (or for that matter his student Jung who developed a different naming of the psyche) But that does not mean we cannot understand the framework of his thinking or endeavor to change the operating system we currently use in viewing and understanding our nature and our condition.

There are many practices. There is meditation. There is peyote. There is walking in the desert. But it may be that the super ego of Freud will soon give way to new horizons- new horizons of allowable thought and behavior. This new collective consciousness will allow us to change individually. Then will come the new ego and sense of self and the new id of motivations.

So the practice is to accept the condition that is upon us now.

Visit with your friends and find those who know. But do not think that it is a waste of time to be with those who do not care or even wish to know. They too are part of the oneness.

Forget about the consumer world that surrounds you. It will exist without your support.

And most importantly, believe that a change is going to come.

BELIEVING

– Believing –

And in all things, whatever ye shall ask in prayer, believing, ye shall receive.

When the Lord had come into the house, the blind men came to him; and he said unto them, "Believe ye that I am able to do this?" They said unto him, Yea, Lord. Then he touched their eyes, saying, "According to your faith be it unto you."

Perhaps the most prevalent understanding of our consciousness is that we must believe. We must believe in God and his story to go to heaven.

We must believe what our eyes tell us. Our lovers must believe in our love for them. We must believe that we can win, that we will prevail, that it's going to be all right. We must believe in our selves, in our country, and in our view of the way things are. We must believe in magic.

Every fairy tale of wonder involves believing. Whether it is a magic kiss for the frog or awakening a sleeping princess, belief is the foundation of the magic. Ask any coach. Any team can beat almost any team on a given day if they believe they can do it.

Someone told me last week during a luncheon meeting that we achieve by setting goals. We must believe that we can achieve the goals and then, much more often than not, we achieve them. The statement, although not meant to be profound, seemed very much so to me. Perhaps I was in my Chauncey Gardner mind.

We believed that we could go to the moon. And we did go to moon. We just stopped going. Who would have believed that? We believed we could stop small pox. Now, except for a few lone batches of pathogens, the disease has been removed from our lives.

Imagine several thousand years ago when the leaders of the tribe imagined that someday everyone would be able to know symbols which when combined together, created abstractions for the mind which allowed for the transmission of knowledge, both mundane and sacred.

Imagine.

What an idea! The utterances of man in search of food and shelter, become transformed into ideas of god and existence, and the meaning of life. With these abstractions, these words, these names, these associations of words, and the finer degrees of discernment, information and knowledge is coded so it can be repeated. Then, imagine a technological innovation such as a pen and paper. What a marked improvement this would be over paints on rock. Imagine keeping these markings on paper in a dry, dark place so that generations could someday access their meanings.

They imagined and ultimately somebody believed.

Now reading and writing is basic to human consciousness.

True, there are many parts of the world where reading is not commonplace, but they are more and more remote. Not only is reading and writing basic, but so is the ability to count and the notion of trade. Humankind has been trading beads for meat and bread long before we were trading novels. If someone were to speak of computers and song sharing and e mail to a Socrates 2500 years ago, it would have been very foreign and very much outside of his ability to imagine or believe. It is the same today in certain remote parts of the earth.

Yet, in this time, the bulk of humanity has evolved into a different kind of creature. The most advanced humankind creature of Socrates' times might not understand the reading of writing on paper to stimulate the desire to get a new car. But they would understand reading the writings of their teachers to more fully understand the nature of the universe they inhabited. But these advanced beings were few and far between. Today, most graduates from junior high can think as well as the best thinkers 2500 years ago.

This is not to say that there are not serious shortcomings in our education establishment and our thinking processes, but the point is this; humankind has evolved and is evolving. There are hundreds of millions of humans on this planet that are well-educated somewhat civilized beings. Today there are hundreds of world-class museums and hundreds of thousands of Alexander quality libraries. Twenty five hundred years ago, there was barely a handful. Today, there is a large mass of developed consciousness that is ready to evolve to a new operating system, with new super-ego hard drive capacity. Not only is our consciousness ready, we now have the technological metaphors to map the way. Twenty years ago, only a few geeks really understood what an operating system was. Ten years ago, only a few really understood what RAM meant in terms of allowing that operating system to run efficiently and quickly. As far as memory is concerned, there is more memory in my cell phone than in the computer that took man to the moon.

All of this and more add up to a belief that the consciousness of humankind is on the verge of a major upgrade.

However, there is a shadow side to this human-computer marriage. Soon, if not already, there will be super smart human beings aided by their cybernetic brain tools that will dwarf the humans around them. Not unlike the way Socrates towered over the slaves who kept the city of Athens running, these new cyber-beings will be capable of processing information and using it to their advantage beyond the capabilities of the masses around them. The coming age of cyber-beings capable of almost super human intelligence is literally around the corner. Just like the advent of nuclear weapons brought and continues to bring special challenges, the coming convergence of human brain power with computer memory and work potential will bring about special challenges for democracy, freedom, and justice.

Perhaps this experiment conducted in 2003 demonstrates the potential and the remarkable capabilities that will be realized in the very near future. Scientists connected electrodes into the brains of monkeys and then set up the monkeys where they could employ a joystick to run a computer controlled image on a video screen. If the monkey ran the image through the right set of situations, it would be rewarded with food. As the monkeys learned to accomplish the modest mental task of moving the joystick to get an image to accomplish a mission, a computer recorded their brain patterns. After a while, the scientists allowed only the brain patterns to actually control the movement of the image on the screen. The joystick became meaningless. Of course, no one told the monkeys. The big surprise came when the monkeys figured it out. Soon, they were manipulating the image on the screen with their minds and not their hands.

Primates have thus accomplished the control of physical things via direct computer- brain interface. There are many more examples occurring on a regular basis in the neurological labs and universities around the world.

This means that in the near future, Tuesday in geologic terms, the capacity and the capabilities of the human brain and consequently human consciousness will be greatly magnified. But it also means that if we are not wise now in our approach to the Holy Field, we will have the opportunity to be stupendously unwise in our approach to reality, our brethren, and to our purpose in this wonderful play of energy and time.

It does not escape me that I have made a judgment here.

Perhaps a continuation of our subject object bicameral mind of right and wrong, good and bad, night and day, alive and dead, awake and asleep, and with or against us, once supercharged by our tools of the mind and its interface with our other tools of control and creation, will lead to a more refined consciousness of harmony and beauty. Perhaps the nail that falls out of the Pentecostal preacher's truck bed which causes the flat, which makes the old man swerve, which causes the crash, that kills the children, that saves the son, who misses the free throw, that lionizes the deserving coach will remind me of the ways of the Holy Field and this magic show we call the creation.

So I must believe.

I must believe that We are enlightened beings whose very nature is transcendence and oneness. I must believe that each of us is capable of revealing the unrevealable, if We would only pay attention.

I must believe that the present is the present. I will remind myself that each moment that we live is a gift. And once I believe it with understanding, I will fully inhabit that moment, knowing that, as Einstein said, that "the future and the past is nothing but a rather persistent illusion."

I must understand that I live in the fields of the mind and within its abstractions and representations of the energy field I dance within. I will try to place this understanding in context to this Holy Field that my mind is trying to explain and describe. And, I will find humor in the misplaced importance I will place on these abstractions of truth. I will use judgment wisely and as a tool of the mind, not a way of being.

I must center myself in the heart of my own universe and find and know my place on the earth, in the solar system, and in the galaxy in this show of energy and time. I will see the sun like a child on a merry-go-round sees his mother go by. She is not setting or rising, for I am simply turning. I will see the moon as a very close friend who travels around me every month. I will know the planets and see that they live on a very large dance floor. I will know where we all are and know that we are spiraling through space at a remarkable speed.

I must honor and respect the naming tool that serves as the foundation of my understanding. I will use it with respect and care. I will know deeply that to name something is to in some degree to take part of its life from it and take it with me. I will know that in this naming, I will likely forget to truly observe what I have named.

I must see and hear and feel and taste and smell and otherwise sense in any way that I can the nature of this Holy Field with a freshness and vitality that enables me to connect to it in the best of my understanding and place in the grand adventure.

I must practice being the being I am. As Bucky Fuller said, I must be a verb. I will not be lazy as I go through this spectacle. I will greet the day as I would greet a new best friend or an old dear lost lover or estranged son who has come home. I will be the best that I can be.

And I must believe.

I must believe that we can go beyond these words, beyond these tiny abstractions of the truth, beyond the veil of the mundane into the rich texture of the sacred in our lives.

I must believe that, as Krishnamurti and many others have said, that the "observer is the observed." When I forget, and start to think otherwise, I will remember the proof of the scientific reality of this statement by a logistician named G Spencer Brown in his book, The Laws of Form. And I will remember the Heisenberg uncertainty principle which states that you can not observe a thing without in someway changing that thing.

Yet, do I believe?

A small, yet growing part of me does believe.

That small part of me says, the more you believe that it is all magic, that this truly is a remarkable display of time and energy, the more magic there will be, and the more magic you will weave, and the more magic you will see, and so on and so on.

It is the most positive of the positive feedback process.

Surely you know what I mean?

Surely you too have become a little suspect of this universe and the way it behaves. How many times have you met a friend or lover in the middle of large city without knowing how you managed to be at the same place at the same time? How many times have you thought to yourself, "I don't know how this happened, but I'm glad it did, and it certainly came just in the knick of time."

Since the "observer is the observed," then what you see, and sense, and feel "is you." It is not the "not you." it is a "reflected you," perhaps more like a mirror, but ever so more subtle.

It is hard to believe that the universe, the Holy Field that you observe is a unique creation of you, not some random disordered serendipity. Yet, serendipity it is. It is ordered chaos.

The more we believe in the Holy Field and know it fully and completely, the more we know ourselves fully and completely. For in all ways, it is one. It is one divided, and it is one undivided. It is much like a hologram. If you break a hologram, you have small versions of the same picture. Division then, is an invention of the mind.

Stacey just told me that the toilet is not working well. I need to go get a plunger. I do not know the word for it in Spanish.

I fix the toilet, and a nice rain comes. We move from the open terrace to the covered terrace, to the vestibule, to inside. The light rain stops. It is cool. The birds fly. The church bell rings three times. In fifteen minutes, it will ring four times, then pause and then ring three times again. It will be three o' clock.

Maya is awake.

The Rising

Just about forty years ago, I was in a rock and roll copy band. We had a few songs that we wrote, but we mostly copied the Beatles, the Rolling Stones, and to my dismay, Paul Revere and the Raiders. We were pretty good for youngsters though.

One late afternoon, the drummer and I went to check out the set up at the Youth Center where we would be playing later in the evening. Satisfied that the set up was under control, we left. There is not a lot to do in the majestic flat lands of the upper high plains of Texas. Almost always, you are pulled to the sky. They are so big there. The eyes and the spirit seemingly have no other place to reside.

Smisson and I had some time to kill, so we decided to buy some 10 cent gliders at the convenience store and go down to the park that was in front of our houses. Smisson was a neighbor and his name really was Smisson. His middle name was Mulkey, and just for the record, he was the third.

Anyway, we took our gliders out of their plastic wrappings and assembled them. There are really only four pieces. There is the main fuselage, the wing that slips through the slit in the fuselage, the back wings, and the rudder that slips on the top of the fuselage.

I put mine together quickly and fine-tuned the wings and rudder. As I did, I said to Smisson, "Watch this Smisson, it will never come down!" I pointed the glider into the ever constant Panhandle north wind and sent it off.

It did what all gliders do. It went down, picked up speed and lift and then went up and over up side down and started to head for the ground where it actually, (and I do remember this faithfully) touched the very top of a delicate wild flower stalk. Then it picked up speed and lift and begin to head back up again. This time, instead of going over backward, it stalled for just a moment and began to dive back towards the ground. As it began to head for the ground it picked up speed and began to climb again, and this time, it climbed a little higher than the time before. Then it stalled. Then it accelerated toward the ground and then turned up and ascended even higher. Then it began to dive and pick up speed again. By this time, Smisson began to pay attention. In the 10 cent glider world, a 10 second ride is pretty rewarding. That would mean that the glider went forward and down, then up and over, then up again, and then peel off to the right or left out of the wind and circle around finally crashing as it began to lose its lift.

But this glider seemed to have a mind of its own. It kept its nose into the wind. It continued to stall, descend and pick up speed and lift, then turn its nose up and move higher into the sky.

I watched.

It went higher and higher.

It kept climbing.

Smisson began to laugh. Soon, I was laughing with some kind of ecstatic joy. By now, the glider was several hundred feet in the air and it had worked it way north of the park towards the street. We followed it, laughing and looking at each other like we were in the first act of pretty good space movie.

We got in the car and followed it as it continued to stall dive and climb, stall dive and climb, stall dive and climb.

We followed it to the north side of town before we completely lost site of it. The last time we saw it, it was still climbing.

Like I told Smisson, "Watch this Smisson, it will never come down."

It never did.

Smisson and I made a deal that day. We decided that no one would believe this story, so we agreed to tell no one. We went back and played that night with that certain kind of smile that you generally get from doing something wrong. And, we never spoke of it again.

About 25 years ago, I was talking to a friend of mine one night and said to him, "Once I threw a glider into the sky and it never came down." The moment I said it I thought to myself, "Did this really happen?" It's the kind of statement that can really damage your credibility even amongst the best of pals in the best of liberal thinking and creative conditions.

I thought and thought about it that night. Smisson would know.

The next morning, I called another high school chum who might know where Smisson was. He told me that my old neighborhood pal was in Amarillo. He was managing a store called Just Boots. (that part may be made up) By that afternoon I had successfully flown from Austin to Amarillo and rented a car. Just Boots was in the west side of town.

I walked into the store. Secretly I spotted Smisson. Like a hunter, I moved in. Our eyes met and a warm smile of recognition is shared. We hug.

"I'm here for a reason," I said.

"What?"

"Do you remember anything about a glider?"

Smisson began to cover his head with his hands as if he didn't want to hear, "Oh no, the glider, the glider, the glider."

"We followed it out of town right?"

"And it disappeared right?"

"It never came down?"

"It never came down."

I learned a lot that day in the park on the High Plains.

I learned that sometimes, the forces line up and unexpected miraculous things occur. I realized that what goes up, doesn't have to come down. And, I began to appreciate just how important words are.

The phone rings. It is Martha. It's after eleven on Sunday night. She sends me a kiss. I tell her that I am just finishing the first book of *Beyond*.

And I am.

Epilogue

In the beginning of this book, I stated that this book is not about religion. That it is not intended to conflict with your belief in God.

I believe in God.

This book is not meant to make you drop your belief system in which Jesus is God. In fact, many of the sayings of our Lord are used in these chapters. I said that this book is not meant to make you question your belief that Jesus is or is not the Messiah and that God is Jehovah. In fact, as you may have glistened from these words, I am a Christian. When I go to the old Church in Real de Catorce and slip back into the small chapel with that particularly tortured Jesus hung on that cross, I pray to him. This book does not deny the Blood of Christ and his sacrifice that we may be free.

If you are a follower of Islam, the book is not meant to make you discontinue your obedience to Allah and the teachings of his prophets.

Equally, your faith in the teachings of the Buddha should not be undermined. Many of the words in these chapters reflect the gifts of the Buddha and his four noble truths. In many ways, the revised operating system that I am offering is Buddhism.

Whether you are a Zen Buddhist or a Full Gospel tongue talking Pentecostal lover of Jesus, you should not find this book to be antagonistic.

And even though I quote many Shamans and their beliefs and teachings about the second sight, this is not the way of the Brujo. These beliefs in consciousness of the spirit soul are not to be confused with the words in this book.

This is not a Way.

Rather, this book is more about learning to read. Or to count. This is not mysticism. This is learning. This is where we are. It is our place and our destiny.

This book is about understanding and applying a new understanding as we participate in the world of space and time. It is not about moving towards the Light and away from Darkness. It is not about good and evil. It is not about the Archangel Michael slaying the dragon that great Satan. It is not about leaving Hell and going to Heaven. It is not about the two lands that are in festivity. It is not about incarnation, astrology, phrenology, or pharmacology. It is not about the end of the World or the return of the Christ or the Resurrection of the Saints.

It is about the end of your world. It is about the end of a consciousness that has served us well and now, like a good plow horse, must be put out to pasture. It is a consciousness that is beyond light and dark.

I believe that we must change the way we think.

I believe that we must change the way we see, and react, and behave.

We have empowered ourselves with weapons of destruction that would make Shiva blush. If we do not change our operating system and bring it up to par with our technologies, the mismatch may bring grave consequences.

It's as if we are trying to pilot a space ship with the mind of the hunter-gatherer.

This is not religion. This is intelligence.

BOOK TWO

Introduction To Book Two

Book Two is not like Book One.

Book One proclaims a new operating system which attempts to point the way to a way of thinking that goes beyond black and white, good and bad, and light and dark.

It purports to offer this new operating system for the human bio-computer in such a way that it does not conflict with your existing beliefs in God, be they Christian or otherwise. It speaks to a way of living and seeing in which judgment is reserved and observation is attuned. It speaks to a way of being in which the "now" is found to be the only true living moment. Yet, it holds that this is just learning; and not unlike counting or abstracting. It holds that the human capacity to be in the oneness is our most natural state and not a state which runs from our being like a mirage on the hot desert highway, always enticing us to come farther only to recede as we approach.

It speaks to a change of consciousness that is available to almost all of us should we choose to embrace it.

Book Two is not so easy.

It challenges all that we know. It cannot and does not blithely state that a change of consciousness does not change everything, including our ideas of God and Politics and Learning itself.

Book Two leans heavily on this developing new consciousness of transcendence as it bores full throttle into this cloud of unknowing.

Book Two charges into our understandings of spirit, consciousness, our walk in space-time, and into this Great Mystery of existence that should undo any pride we may have accumulated along the way as a result of our teeming baskets full of gathered mind flowers.

– Beyond –

Like Book One, Book Two is also not a Way. Your way still may be through Jesus or Buddha or Allah. Jesus said no man cometh unto the Father except through me. Or your way may simply be your Way.

Rather, these are the words of the Traveler, one who like you, walks through the mystery of the Holy Field with the determination of a miner and the delight of a child.

I start with the Wind, for the Wind gives me Hope.

.

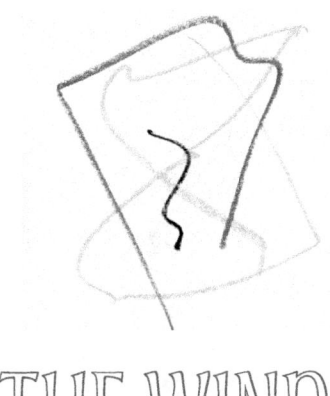

THE WIND

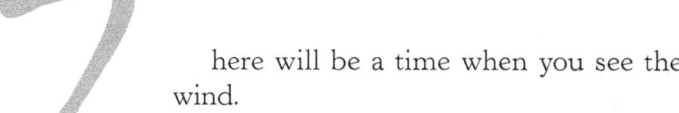here will be a time when you see the wind.

You will see it in the trees.

You will see it in the grass that waves as the wind rushes across it.

You will see it on the water.

You will see it in the heavens.

No longer will it be invisible.

And you will see that the trees make the wind blow

as surely as the waves do.

You will see that the grass touches the wind.

You will see that it is all one.

And you will see the perfection of it and the marvel of it all.

Breathing In the Spirit

The air we breathe is pretty much a miracle.

Most of it is nitrogen. Nitrogen is a relatively inert gas. Except of course when you are talking nitro-glycerin or a bomb made with nitrogen fertilizer. Then, it sounds more like a really dangerous element. We breathe it in and out every day. We breathe it so we can get the oxygen. That is the stuff that allows our cells to live.

Although obvious, it is worth noting.

I can go for weeks without food and days without water, and months without love, but I cannot exist for more than a few moments without oxygen.

Think about it. We all know from the time we are tiny children that we have to breathe.

Suffocation is an early concept. "You could be suffocated by that pillow," or that plastic bag, or under water, or by your own car in the garage, or heaven forbid, in a fire.

Our breath is our life.

And as long as we are breathing, we are still alive.

"Is he breathing?"

"Yes, I think so."

So he is alive.

Everyday, I breathe maybe 15 to 20 times a minute. That's 1000 breaths an hour, and 24,000 breaths a day. If you're lucky, you may breathe almost a billion breaths in your short walk on this earth.

But we breathe unconsciously. If we had to breathe consciously, it would be more like a job. But we don't have to think about it.

And in the mind that can see that the trees make the wind blow, then perhaps we can imagine that we are being breathed just as much as we are breathing. Perhaps the animating gas we breathe is animated itself. To see that the trees make the wind and to know that you are being breathed, frees the mind and the heart. It connects you to all that you are and to all that there is.

The wind, the breath, the spirit…. they are one.

— *The Wind* —

So in this Holy Field of energy and time, there is another force that resides within this field. It more than resides in it, it permeates it. This force is spirit.

We know about spirit. We were taught early about school spirit and how to go to things called pep rallies to whip up our spirit. There are the spirits in a bottle of whiskey. That is why they are called spirits. There are days when we are in good spirits. There are times when are spirits are low. We know that some places have a good spirit. We know that when we are with certain people, our spirits soar.

We all know of spirits.

The Bible speaks of the Holy Spirit. This Spirit is part of the three-fold nature of God as the Father, the Son, and the Holy Spirit.

The Apostles waited for the Holy Spirit on the Day of Pentecost. It is this spirit that empowers the Church. When the Lord was baptized by John the Baptist, a dove flew down from heaven as a sign of the Holy Spirit.

In one way or another, everyone who prays, prays to the spirit. The Indians of North America prayed to the Great Spirit. Christians pray to God in the name of the Holy Spirit. Shamans see and interact with the spirit.

For it is through the spirit that we see the Father and the Son. And it is through the Spirit that we see that they are indeed One. It is through the spirit that we gain additional access to the Holy Field of Space and Time. The world of Spirit is like another TV channel. Or perhaps more accurately, it is more like a new satellite dish that gives you a completely different menu of selections in which you can view the Holy Field.

Last night, over dinner at the Abundancia Hotel, the three ladies that were at the end of the table began to speak of ghosts in their old hotel rooms. The daughter of the retired high school principal said that there was a ghost in the bottom floors of the hotel we were dining in. And, apparently, Humberto, my friend who has another Hotel up the street, has a ghost in his hotel too. This should come as no surprise. The Pueblo is 250 years old and given that it is an old mining and minting town, it has surely seen more than its fair share of greed, desire, and violence.

Then the conversation leaped to the admission that the younger woman would see lights in the corners of the room that would sometimes "come together" as if they were playful children. And then, there was the story about the lady who could not keep her hairpieces because they would all mysteriously disappear. (and she didn't have pets) This encouraged the other lady who was traveling with the retired school principal to talk of some spirit who would caress her neck while in the shower of some apartment she had rented. She went on that the landlord had warned her and her brother had confirmed to her that she had "spirits" up there. The good news was that they were harmless. She recounted that with time she even enjoyed the caresses in the shower. And she was an engineer!

Some speak of spirits as if they eat breakfast with them every morning. Maybe they do.

Others speak of spirits as anomalies to the world of physics. The stories they tell are unexplainable. Yet, they know it happened. They saw it. They heard it.

It is ironic, that many of us pray to some kind of spirit, yet our western trained brains do not accept the existence of such a thing. It is no wonder then that faith is stressed in the Christian tradition.

Was it a spirit that lifted the glider from my hand to the heavens?

Or was it the equivalent coincidence of throwing a nickel on the floor to see it land on its edge? And is there a difference?

— *The Wind* —

The Yogis speak of something in the air besides oxygen called prana. Indeed, Pranayama is breathing in this "life force" as a form of meditation and health. An entire school of Yoga is based on Breathing.

We are told to breathe before we speak in public.

We remind ourselves to breathe when we are about to lose it in the traffic jam.

We take a breath when we make our mind up and become determined.

Every swimmer knows to take a deep breath before you dive into the deep end of the swimming pool.

We protect our breath as if it is our lives, because it is.

It is the Spirit in the Wind.

The days here in Real de Catorce in the summer time are generally almost perfect. They are mild with strong sun. In the afternoons, the storm that was brewing on the mountains may decide to come into town through one valley or another. We are then blessed with a nice rain that cleans the streets and the air. Often, in the late evenings though, a strong wind arises. Almost every night is full of viento fuerta. This invisible force rattles everything. It makes the roof shingles flap and the windows rattle. It blows over and spills the bottle full of roses on the table outside. This morning, the chairs outside had moved and arranged themselves in their own wind picked positions. This invisible force comes through the Pueblo and rearranges chairs, dust, and rocks and then leaves every morning so the birds can sing their good mornings.

It is no different with Spirit.

It blows into our lives often at night and rearranges our lives. It knocks over our carefully arranged flowers and our ideas of how things should be. Although invisible, it flaps the roof of our understanding and it rattles the windows of our perception.

Yet for most of us, we simply go out in the morning and straighten things up.

When the invisible wind of the Spirit comes into our lives, we seek to repair the damage and return our life to where it was. Yet we know the wind will come again, and again, and again. Or, we begin to learn to listen to the spirit. We begin to learn to see it in our lives. We see the spirit in our lives just as we see the wind in the trees. We see the spirit in the waves of our consciousness.

Sometimes, I wish the spirit would behave.

But I do not want it to leave me alone.

According to the Bible, the only unpardonable sin is to grieve the Holy Spirit.

Power In the Wind

Many years ago, after I had left college and was just beginning to make a family of my own, I made the long trip back to the high plains of Texas from the Central Texas lowlands that I have made my home.

The drive is almost 600 miles. My father lived in Pampa. Our family had been there for about 100 years. In fact, my Great Grandfather had come to Texas after fighting in the Civil War as a very young man. He survived, but the world he knew did not.

I have always been one to pay attention to the signs. This night, the sign was my odometer. I know it seems odd if not a little frivolous to let odometers run your life. But somehow, with several hundred miles to go in my journey home, it became obvious than the mileage on this white 2 door Ford would hit 77777. And I knew I should pay attention.

As I drove into the late evening, it became more and more obvious that I would hit the 77777.7 mark very close to the outskirts of town. Finally, as the numbers came closer and closer to their alignment, and I saw the lights of town, it became clear. At exactly 77777.7, I stopped the car. There I was, at the corner of the lumber company where I had spent hours as a child. Owned by my father's older brother, Johnson Polk Osborne, Pampa Lumber Company was a pillar of my childhood.

There on the corner was something I had never really noticed before. A Chicago Air-motor, just like the ones you see dotting the Great Plains, towered above me. My family was a distributor and installer for this important piece of machinery that had virtually tamed the west by bringing precious water to the surface to the thirsty herds of cattle so the cowboys and ranchers could actually tame the land.

I stayed there, staring until I got it.

A few years later, I installed the largest electric generating wind farm in Texas on land owned by another Uncle on the north side of town. At the time, it was the second largest installation of wind power in the World.

I stayed in the Wind business and helped find the wind for the companies that followed my lead. I would go out and lease land and install towers with measuring equipment that would document the presence of suitable wind for economic development.

Over time, I now receive Wind royalties for my efforts. And thousands and thousands of people in Texas and the World get some of their electricity from "state of the art" wind turbines that can significantly reduce our use of the fuels that fowl our air and abscond with our capital. There are not many people on the planet who receive money for wind but there will be a lot more as the idea becomes a powerful industry and not a novelty or green jewelry for coal burning utilities.

So, you see, I do know something of the Wind.

There is Power in the Wind.

This Power is not only physical, it is also spiritual.

There is something mystical and terrifying about the Tornado. Here is a wind that travels in a column of twisting, turning air and water vapor. It comes down out of the clouds like the finger of God. It can destroy a community or it can mysteriously place a stick in the trunk of an oak tree. It can lift the house off of its foundations and leave the owners holding on to the plumbing in the bathroom.

The Whirlwind came and took Isaiah to Heaven. He was the second to go without knowing death. We all have known the Whirlwind.

This is the time, that the Spirit speaks in a way that cannot be ignored. You can find your life uprooted and your way of being totally changed in the proverbial twinkling of an eye. It can be that bad car wreck or the untimely death of your wife. It can come in the form of disease which forever takes of your health. Maybe the whirlwind that comes into your life can be forgotten, but it cannot be ignored.

When the spiritual Whirlwind comes into your life, be grateful, not fearful. For it is a time of change and of a quickening of the spirit. You will survive. Or you will not. Either way, you have no choice in the matter. Like Isaiah, you may be plucked from this plane and taken to another.

Or like Dorothy, you may be whisked to the Land of Oz, where you will meet three friends who will walk with you to see the Wizard. You may be required to slay a wicked witch. You may have another good witch who will watch over you. Your friends will seek courage, wisdom, and a heart. But you will want only to go home. But truly, after the work of the whirlwind in our lives, we can never go home. We can, at best, find that we never left. But more likely than not, we will find what my great grandfather found when he returned to his family's homestead after the American Civil War. We will find that what we knew and what we loved no longer exists as it did before. We realize deeply that if we are to move on with our lives, we must leave the past in our lives.

As the novelist said, it was "gone with the wind."

The Whirlwind Of the Spirit

One day, you may find that you must meet the Great Whirlwind. When you see it, run. But don't run away. Run into it.

Many years ago, I dreamed of a whirlwind. Within that Whirlwind, there was a man, a Master. Upon seeing him in the Whirlwind, I ran into the twirling cloud.

Immediately, I was lifted and my spirit was quickened. A voice came upon me. My partner awoke from her sleep just before. She said it was as if time stopped and everything became very still. I briefly awoke to the power of the spirit.

A few years later, I was hitchhiking from Texas to California. I was driving a yellow import Buick Opel for the guy who had picked me up. He was driving from New Jersey to Los Angeles because he had been given a new opportunity with the grocery chain he worked for.

As we came down one of those long Mojave Desert Hills and began the 15 mile climb up the next rise, I saw a dust devil on the horizon. It was miles away. I was driving and my weary traveler had fallen asleep. Only hours before, I had been held by the police in Kingman, Arizona. I had been traveling all night with a young escaped convict and his girlfriend. They too, had asked me to drive. When they dropped me off in Kingman to head north to Las Vegas, they apparently were picked up by the police just outside of town. And, he apparently told the police that I was a dangerous criminal too. As my New Jersey lift drove by, he saw a bearded hippy with his arms stretched out on a police car and his legs spread eagle. I saw his eyes as he passed and could see that he had turned into the convenience store across the street.

He then got out, cupped his hands to make a megaphone around his mouth, and yelled, "Hey buddy, do you want a ride?"

I affirmed that I did, as soon as this little problem was resolved. The officer heard the man, and as it became more than clear that I was a soul traveler and not a murderer, he told me to catch that ride and never return to Kingman. I never have.

So, my New Jersey man and I are now headed for L A and I see this Whirlwind on the horizon. It is a towering swirl of dust and debris reaching hundred of feet into the desert sky and it is clear to me that it is coming. As I maintain my speed, I marvel at how this whirlwind seemed to have some kind of radar to track my progress up the long hill. I watched it and watched it. Then at the top of the hill, it crossed the highway like a freight train exactly as we came to the top. It entered the car and blew all the papers around. It made the car go a little sideways. Then, just as suddenly as it came, it was gone.

"What was that"?

"A Whirlwind," I answered.

"Are we alright?"

"I think so."

Now I was going to LA to see a Holy Man. And when I got there. He and his Ashram had moved to San Francisco.

When I got to San Francisco, I was told that I could not see the Holy Man until I had spent time (and money) sitting with the group. But there on the desk, was a rolodex with the Master's home phone number. As I watched the movements of the mouth of the man who was telling me that I could not see the Master, I memorized the number that was on the rolodex card before me.

I walked out of the Ashram and I checked into a hotel just down from the Top of the Mark. It was the Commodore Hotel. I thought and thought about calling him and what I would say if he answered.

I never dialed the number.

The next day I flew back to Texas on a plane called the Golden Eagle. On the hitchhike from Dallas, I noticed a white station wagon like vehicle coming down the Interstate. When you are hitchhiking, you generally know which ride is yours. The vehicle came closer and I looked more and more closely at it. It slowed down and stopped in front of me. The heavily tinted electric windows came down with power motor precision. Inside at the wheel sat a thin man with dark sunglasses and a black suit.

"Want a lift?" he says.

I look into the rear of the vehicle at the casket. "He's dead", the driver says.

I get in.

And I take my ride home.

A few weeks later, that Master that I had gone so far to visit, came to me instead. Traveling in the soul world, he brought an explosion of ideas, creativity, and insight into my dream world. Such is the way of a Jnana Yogi. I remember feeling and thinking that my mind had literally been opened to the light of day.

The next day, one of my housemates asked if there had been a visitor in the night. He had been awakened by what he thought was a guest at my door.

I told him that, in fact, I did have a guest during the night.

He said, "but it was that Guru guy you went to see. Was he here?"

I nodded. He just looked at me. We agreed to keep it between ourselves.

I have not looked for the Guru since.

For the presence of God is everywhere.

I wonder what God thinks when he sees someone looking for him. (or her)

He must be flattered. And a little confused.

It's a little like someone who closes their eyes and says, "I can't see."

Or perhaps, it's more like learning to see the Wind.

THE SEA

— The Sea —

I thought I was going to the mountains to my home high in the Sierra Catorce.

But I went left in San Antonio at the junction of 37 and 35 and came to the old fishing town where Franklin Deleanor Roosevelt used to fish for Tarpon in his early presidency. In fact, I am staying in the same Hotel that the President used to house his party, The Tarpon Inn. It is 120 years old, so says the Texas historical plaque on the sign in the front.

I checked in and made my way to the beach well before sunset. Now, there are more than a few who will belittle our beaches here in Texas, but on the right day, in the right season, they are perfect. The water is blue-green, the sand is just right for castle building and long barefoot walks, and the sea weed that is left on the crème colored sand by the tide is that odd yellow and red cord from the Sargasso Sea. And this being a Sunday night, the crowds are gone and the beach is almost sans sapien.

I walked out to the water and touched it. It was very warm. The Gulf of Mexico is very warm now with climate change. The birds here are part of the spectacle. Why the pelicans make their sorties in groups of three I don't know, but it makes for wonderful hotel and condo art.

I saluted the Sea and took a picture of it with my cell phone. A Pelican flew into the center of the photo just as I took it, as if to say, "Hey wait, you need the single pelican shot." I gathered and centered. I listened to the water and the waves. I heard the birds.

I looked at the magnificent sky that was opening to the east and then saw the full moon rising as the sun hid beneath the earth behind me. I feel the vibration of the full spectrum of energy here. I walk to the car and get in after shaking the fine sand out of the grooves of my New Balance shoes. I start the car.

The mileage odometer says 73,000 exactly to that point. I figure I made the right decision. Besides, how can you write a chapter on the Sea in the Mountains?

It took me years to learn to totally love the beach. I would always get sunburned and bored. Then, I learned to bring a chair, a mat, a cooler, an umbrella, a full tube of sun block, and that book I'd been wanting to read. I read, watch the birds, the babies, the people, the waves, the sky, the changing tides and landscape during the day while I listen to the sounds of the never ending waves coupled with the seemingly insane laughs of sea gulls in concert with occasional blasts from afar. I find that the sound of the constant washing of the waves onto the beach somehow reprograms the subconscious mind. But it is a slow process. You must plan to spend at least 3 days or 24 hours washing your subconscious. By the end of the day, you are tired but refreshed.

Water World

If the air we breathe is pretty much a miracle, so is the water that covers most of our host planet. Where I can only live for a few minutes without air, I can only live a few days without water. I am a water-being actually. Most of the body weight is water. We are like jelly bodies on a calcium frame with a very thin glass jar around us.

And our blood is very much like sea water. It's as if we decided to walk out of the ocean one day, taking part of it with us.

When we look for life on other planets, we look first for an atmosphere and then we look for water. (Why we presume that other life must be like our life seems a little silly but then we are a little silly). When we look to Mars, we see what looks like the results of water passing over its plains and valleys giving rise to our suspicions that there may have been water there at some time.

Water is actually oxidized (or burned) hydrogen. It is a wonderful bipolar molecule that in the world of chemistry makes for a very handy substance. It freezes at just the right temperature and when it does, in violation of all the known rules of science, it expands. That is a good thing for fish living underneath frozen rivers. It is a bad thing for landlords who have to fix their broken pipes when tenants leave for Christmas and they forget to leave their heat on. Pure water is an excellent solvent. Hence, water is useful in bathing. Water with CO_2 makes for a light acid (carbonic acid) that, over time, can weather rocks and earth to wonderful water worn features. Increased CO_2 from the burning of fossil fuels mixing with the water in the air may accelerate this geographically slow acting yet powerful effect of nature.

Water mixes with about everything except oil, hence the oil and water metaphor. This special quality allows the weeks of built up oil on the road to rise up when it rains, making the road much slicker that you would imagine when it is a light rain after a long dry spell.

Water in the Sea has salt in it; water in our lakes does not. This makes for two distinct worlds of water, fresh water and sea water. Where they meet, there is this confluence of worlds that engenders another world that fishermen love. It is at this edge that different possibilities exist.

The Waters Edge

One reason I love the beach so is the way that thousands and thousands come to the waters edge to get close to it. Today, I saw a small family laying at the waters edge playing in the ooze with the water lapping about as their small child sat in the water next to his parents. They all just kind of hung out right at the edge playing and looking for whatever. What is remarkable is how comfortable most babies are in the ocean. If they see that mom and dad are comfortable, they seem to naturally take to the environment. At our Texas beaches, there are man of war, jelly fish, sharks, crabs, and even flesh eating bacteria, not to mention the sometimes tricky undertow and the always possible specter of death by drowning simply by swimming a little to far out into the ocean and getting pulled out to sea by currents. Babies seem to survive it somehow.

If you were viewing from outer space, you would note that millions of the virus like humans who are not capable of living without their symbiotic existence with the primary being of the planet, (a rectangular being with glowing eyes and four round feet), occasionally come out of their host as much as several hundred feet and make their way down to the edge where the water and the land mass meet. These virus like beings (if indeed they are alive and not just code looking for a host) seem to move in huge droves to these edges during certain dates which correspond roughly to the dates in which the days and the nights are becoming equal. We would speculate that the large movement was perhaps a time for mating if indeed these creatures did such a thing. They might just reproduce by dividing.

We know better. I think.

We know that people crave the edge. Most of the people on earth live within a few hundred miles of the edge. At the edge, we are one with all of the elements. We have the water, the earth, the wind, and the fire of the sun. And we have the metaphor of our existence. We have the constant waves hitting the beach. We have the wind coming in during the day and going out during the night. We have the tide changing every six hours or so. We walk without shoes. We dress in painted underwear. We walk around almost naked with hardly a thought of how undressed we really are.

Here at the edge, we play with kites and we throw round things to each other. Some build castles and truly substantial sculptures that will last only till the next tide or the unending wear of wind and rain. But for all of the things that are done on the edge, the most prevalent activity is the no-thing. At the edge, you can do nothing for hours and days at a time. Some say that they are trying to bake color into their skin, but that seems to be more of an excuse and a ritual than a real purpose.

But the edge is more than just this. For thousands of years, humans have come to the Sea for food. The Sea, for many cultures, is the primary source of nourishment. Some island cultures have very little agriculture in their history. Here at the edge are fishermen. Some may work like manufacturer workers on large boats that use advanced techniques to harvest the ocean very efficiently. Other fishermen many simply be exotic tour guides for curious tourists. Some fishermen still go out and catch fish pretty much the same way they did 2000 years ago. But, industrial fishing fleets are good at what they do, and because they are so good at doing what they do, the Sea is beginning to be over fished according to a recent United Nations report on the State of the Oceans.

For many Sea cultures, the Sea represents life. It is as strong a symbol for life as the sun itself. Some inland cultures do not have this same respect for the Sea, but the image of the Sea is strong in all cultures. Every child knows of the Sea whether or not they have actually seen it or not. Every child has an understanding of a very large body of water out there somewhere.

I was born a land lubber as the old salts would say. I come from the High Plains of the Panhandle of Texas, almost eight hundred miles from the Sea. The first time I saw the Pacific Ocean, some 1200 miles to the West, something very basic and strong was awakened in my consciousness and in my soul.

The Sea As Metaphor

Our mind works in metaphors. When we speak of a major change in the way we view a subject, it is a "Sea Change." If there are hundreds of thousands of people at a demonstration, it is a "sea of demonstrators."

When we wish to give borders to this nation we sing of "From Sea To Shining Sea." If someone has been everywhere, they are from or have been to the "Seven Seas." When times are tough, you are going through "Stormy Seas."

An American classic of story telling and struggle is the "Old Man and the Sea." Men go to Sea to discover and find out of what they are made. Ahab went to Sea to wrestle with the Great White Whale.

There are few metaphors as strong as the sea.

And there are few forces as strong as the sea.

For the sea literally connects us all.

Perhaps the strongest metaphor of the Sea comes in understanding this connectedness. In psychology, for example, the Sea often represents our collective consciousness. This is often referred to as the collective unconscious.

Since water is a dipole molecule with both a negative and positive charge, it is capable of carrying current. It is also capable of storing current like a capacitor. This makes it unique. It is both a cream rinse and a dessert topping.

If you've ever wondered why your body made the TV signal stronger when you walked closer to it, it's because of the body capacitance from the water in your body. Small impurities from say limestone in pure water add the ions that can make water come alive electrically. Did you ever wonder why calcium builds up on the inside of copper pipes? Adding salt makes water heavier and even more electrically alive.

One of the first experiments in electricity involved measuring voltage from the Tower Bridge in London. The river passing underneath the bridge had a current to it. That means that all the waters of the world are electrically connected. And if all the waters have electricity in them, then all of our waters have an electro-magnetic field. And that field is the scientific home of consciousness and modern communication.

Ever notice how schools of fish seem to be more like a single organism than they are separate entities? They all move with complete union. They put the best synchronous dance swim teams to shame. The fish in the oceans seem to be able to communicate at lighting speed. And perhaps it is.

John Lilly, the famous scientist who worked with the mind of the dolphins and ultimately with his own human biocomputer through the use of sensory deprivation coupled with brain stimulation, was able to communicate quite effectively with the dolphins. However, as he was nearing the end of his studies, he noticed that he was not communicating with the dolphins, they were communicating with him. They would allow themselves to be caught and brought under his control. And when he did decide to end the study, his best student voluntarily decided to end its life after knowing his intentions before he spoke them to anyone.

Dolphins are mammals that live in the water. They may be the smartest mammal species on the planet. Porpoises, dolphins, and whales are closely related

Not long ago while sailing across the Bay to Padre Island, a group of porpoises came up to the starboard side of our small sailing vessel. They are very curious creatures. We had two young girls on board who were not familiar with such large watery creatures. They seemed to know that. So, two of them escorted us within inches of the boat for perhaps 10 minutes. The girls could pet them as we sailed through the water. It seemed like they were inviting us to just jump on and ride with them. (I realize that is a terribly human thing to say). Once the great charm and intoxication with their presence wore off, they were gone.

Since we are all water beings and all of the sea, we are all able to communicate though our watery source. In our mutual composition of the water element, we are truly one. Surely, the cells in the body do not debate their oneness.

The collective consciousness, be it subconscious, unconsciousness or conscious, is often represented by the Sea. Think about it. All of the thoughts ever thought are recorded in the ocean as surely as all the thoughts ever thought are traveling as vibrations of light through the vastness of space time. (including this thought too)

The ocean and the connectiveness of water serves as our combined network drive. Everything that our individual hard drive thinks (which is after all just modulated light) is automatically backed up on the Sea drive. We wonder how our mother and girl friends know what we are thinking.

Please.

The presence of water in our bodies, in the air, in the food, in the rivers, and even in our plumbing, gives us a very strong network of connections to each other, to all of our species, to other species, and to some degree to those who exist now only in this field (the so called dead).

So the metaphor of the Sea is much more than a metaphor.

It is the truth.

The Sea is our collective consciousness.

It is very much the soul of our planet.

The Sea As Planetary Consciousness

There are so many references to something coming up out of the Sea.

This one in Revelation comes to my mind. It starts in Chapter 13 verse 2.

– The Sea –

"Then I stood on the sand of the sea. And I saw a great beast rising up out of the sea, having seven heads and ten horns, and on his horns ten crowns, and on his heads a blasphemous name."

And then in verse 4, "So they worshipped the dragon who gave authority to the beast; and they worshipped the beast, saying "Who is like the beast? Who is able to make war with him?"

This kind of stuff makes Stephen King look like a fairy tale writer for children.

But taken quite literally in the context that all of our thoughts and collective thoughts are recorded in the planetary consciousness of our waters, our Sea, then this big dragon comes from us, at least the political, social part comes from us. (I suppose the Porpoises have their eschatological issues too.) But this dragon comes from the sea, our Sea.

And it our own collective consciousness that provides the basis for this upwelling of events portrayed in the vision of St John.

If indeed, our waters act as an electric circuit, as a giant capacitor, and a long term memory device for all of the collected thoughts of humankind over the eons, then there is little doubt that at some time, all of the hate, all of the bad thinking, all of the greed, all of the military cruelty, all of the violence, all of the jealously, all of the lust, and all of the plain old lower bubba mind stuff that is the hallmark of our short time on the planet will at some time fester up into some upwelling of rather unprecedented proportions. This collected memory, call it Karma if you like, could bring about some serious upheaval.

However, it seems that if we understand our situation, it is a beginning to changing our situation. At least we have made it conscious. Enormous amounts of collected memory can be made to work for us if we chose to make it so.

One of the Books in the Bible that always intrigued me is the Book of Jonah. The Lord came to Jonah and told him to "go to

Nineveh, that great city and cry out against it; for their wickedness has come up before Me."

Jonah instead decided to run to Tarshish. So he got a ticket on a boat. "The Lord sent out a great wind on the sea, and there was a mighty tempest on the sea, so that the ship was about to be broken up."

The Captain figures out that someone on board has a job to do and tells Jonah, "Arise, call on your God." They even cast lots and Jonah came up with the black bean. Jonah then confesses and tells the men to throw him overboard into the sea so that it may be calm for them. The men try hard to row to safety but to no avail and so they throw Jonah overboard as the last resort. A great fish (a whale) then eats Jonah where he stays in its belly for three days and three nights until he is vomited out on dry land.

The Lord tells Jonah, once again, to go to Nineveh. This time, he goes.

When he arrives after three days journey, he says, "Yet forty days and Nineveh shall be overthrown." But the People of Nineveh believed God and they proclaimed a fast and they put on sackcloth. Even the King covered himself in ashes. And the King said, "Who can tell if God will turn and relent, and turn away from His fierce anger, so that we may not perish?"

"Then God saw their works, that they turned from their evil way; and God relented from the disaster that he had said He would bring upon them, and He did not do it."

Now this made Jonah angry. The Lord did not tell him to tell Nineveh to repent, He told Jonah to tell them they were to be destroyed. Jonah decides it would be better to die, so he goes out and lies in the hot sun. Then, the Lord frustrates that by making a plant grow up over his head. The next day, the Lord provides a worm that eats the plant, once again leaving Jonah with his anger and surely thoroughly confused. Jonah's anger came from his pride. He was more concerned with his own pride than he was able to rejoice in the saving of 120,000 persons. I guess nobody ever said that messengers have to be perfect. Maybe that's why so many get shot.

But the point is this. The people of Nineveh saved themselves by believing and by changing. And God changed his mind.

The Beach Head

Yesterday, while power relaxing on the beach, the winds came up beyond their normal afternoon peak. That caused my modern box kite to begin to move to the right and down. (I'm not sure why) For a moment, I thought it might dive down and attack the older man and woman with the deep dark professional beach tans that come only with years and years of power beaching. I always fly my kite over my beach camp. It makes it easy to find when I walk or run down the shoreline from the camp. And, it's just another way to touch the elements. Specifically, I touch the wind.

So, I walked down the kite with the kite string where I made some adjustments on the bridle trying to get the kite fly straight into the wind. When I released it, it flew off in the now gale force winds and pulled out the string spool that was stuck in the sand. I managed to catch it and as I did, the jolt made the cross sticks bow, throwing them out to the ground causing the kite to lose its flying integrity. By now, I was several hundred feet from my beach camp and getting close to the beach road and the houses and occasional beach hotels that face the road behind the dunes.

I then regrouped and put the kite back together. This time when I released it, it took off like a cannon, burning my fingers as I tried to slow it down. I grabbed the quickly running string with my T shirt and once again I gained control. But almost simultaneously, the strong wind bowed the support sticks, moving them out of their stirrups, and the kite blew apart again. This time, it was across the road and in the sand dunes that protect the houses. I watched as each thin cross bar piece fell out of the sky into the groundcover on the dunes. The kite, although still flying, was mostly just pulling, so I eased it down in front of the dunes. Now I needed shoes and a total regrouping. I tied the limp kite off after marking where the sticks had fallen in my head. As I walked back to my beach camp, I saw that my colorful, artful little camp had also been ravaged by the wind. The umbrellas were flailing wildly, the red lounger chairs were turned over, and it was just a mess. So I quickly took the umbrellas down, straightened up things a little, put on my shoes and made my way back to the kite. I started winding up the string as I worked my way back to the kite. But when I got to the beach road, the string was broken, and the kite was gone.

I walked over to the scene and found both cross pieces, but no easy-to-spot kite. I walked up on the walkway that goes up and over the dune expecting to see it. No kite. I looked downwind. No kite. I looked back to the road again. No kite. I looked over at two guys who were standing on a tall wood deck just a few yards away at a rather cheesy, yet local hangout hotel bar. They seemed to be watching me too closely.

"Did you guys see a kite," I yelled into the wind filled air.

"What?"

"Did you see a kite go by?"

"It's on its way to Austin," one of them said.

I walked closer. "Did you really see a kite or are you messing with me", I asked.

Then the one on the right said, "We thought you said your mother was a dike."

His pal joined in, "What color was it?"

Without a blink I responded, "Well my mother is white, but the kite was colored."

Somehow, that was just racial enough for these two early afternoon drunks to find themselves embarrassed. So embarrassed, they both immediately left the deck and walked to their vans and drove off.

I kept looking for the kite, walking deep into the down wind neighborhood areas.

It had simply disappeared.

I talked to the guy with the weed eater who was trimming back the wild ground cover in front of the hotel where the "kite dike" conversation occurred. We both agreed that stealing a kite was pretty weird Karma. Especially one that wasn't going to fly without the special cross rods.

I just don't know where it went. And it's rare that I don't know or feel what has happened. It was downright mystical.

Maybe the wind brought it back aloft into the air and a passing car caught the string and dragged it down the beach in some ignominious way. Maybe it took off and flew and flew and flew in on its way to a new adventure and a new life free of the sameness it had endured with me. Perhaps some deserving child will find it and his father will make some cross sticks for him and they will have a lot of joy flying my former kite.

There is a consciousness on the beach that is personified by the Kite Dike guys. They drink lots of beer. They probably watch Dallas Cowboy football. They are Americans and whatever mascot their high school used (I'm a Harvester), they smoke tobacco and a little pot and they probably used to eat speed. Their hair is that straight sandy blond look that all surfer types have. They are homophobic. They think that just rhyming with the word dike is funny. They might imagine dating a black woman but they would never be able to stand the peer pressure. If they voted, they would vote for the Republicans because Republican think it's cool for you to own your own assault weapons and besides Republican are racist too. (If they knew that Republicans keep the minimum wage low, that would be OK too because they make more than minimum wage)

They don't go to church. They don't read philosophy. They may not even read the sports pages, but they do know about fishing spots and tides. They can read, but they just don't do it that often. And, in all fairness, who can blame them? They live on a beach island.

In the fullness of things, these guys are OK. They probably didn't steal my kite. They are probably pretty decent guys. They wouldn't read my books probably, but they would care about climate change raising the ocean level and submerging Padre Island, thus destroying the life they know and the world they probably even love.

I would guess that these guys are pretty much at the median of human consciousness.

To a guy who is writing a book on a new operating system for the human biocomputer, their presence is a real eye popper, and if I wasn't so sober already, I would say sobering. The water soul of this planet is full of their thoughtforms and emotions.

But our planetary collective consciousness is also full of millions of powerful loving thoughts and emotions too. Even the worst of us can be truly caring and sometimes loving. As bad as we are, most of us have a sense of decency. I've seen beautiful acts of care and kindness come from the most unlikely of places and events. It was the guy trimming the weeds who brought up the concept of Karma and Kites, not me.

So the watery soul of this planet, our collective consciousness, our Freudian super ego on steroids with 6 trillion terabites of memory is more likely in a delicate balance between love and destruction, growth and decline than it is a total ball of negativity.

A lot of people have been putting a lot of love out there for a long time.

Like Nineveh, perhaps humankind needs just 120,000 souls who would listen and change. These souls, like the brave soldiers who fought their way to the beaches of Normandy, could make a beachhead on the shore of our new human consciousness, so that the rest of the army could follow.

Perhaps we are out there now, looking for the right time and the perfect place.

When the wind wrecks your camp and it steals your kite, it may be the time and the place.

We transcend the present way of thinking and behaving. We understand that what might be good for us, is bad for someone else. We recognize that we are all on this planet together and that nation states are worn out concepts from a distant tribal past. We realize that our democratic capitalist system is just a modern extension of the Monarchial oppression that has ruled our affairs and lives through the domination of money for these many centuries. We see that "consumerism" is in fact a wolf in wolf's clothing. We throw away the safety net from death offered by your neighborhood Church of Fear and instead face the angel of death squarely, thus making death our ally and courage our cause.

We understand that our minds are locked up in antiquated thought forms that cannot move us beyond this wall of antiquated code and into the real world of the Holy Field. We come face to face with reality and move through it as Jedhi Warriors feeling, seeing, and being in the Oneness.

We make our beachhead and move inland from there.

I'm going back to the Sea.

For the Sea speaks.

– Beyond –

And the Sea listens.

And I'm going to get another Kite.

THE CREATION

The Creation

Almost every culture has a creation story.

And I'm not going to go into them. Because you see, it doesn't matter. What matters is that there is a Creation.

OK, OK. Well at the beginning, there was this bunny, this big, big bunny.

Everything that exists, all the stars, the galaxies, the sun, the moon, the bad soap operas all of it came from this bunny.

I call it the Big Bunny Theory.

Scientists have even listened deep into space and have heard a distant thump, thump, thump, thus giving more and more credence to the Theory.

Now we've got that out of the way.

I often marvel how people who kind of want to believe in something other than bricks and mortar have the same plea. "Why doesn't God just reveal himself, you know, do some miracles like in the old days. It would be easy to believe in God if there were some people turning into stone, or some guy was healing blind people, or the lame. Or how about turning some more water into wine?"

I know it sounds like one of those feel good books that has those little sayings in it, but it really is true.

– *Beyond* –

The World Is a Miracle

If the air we breathe is a miracle and the water we drink is a miracle, the earth and soil beneath our feet is no exception. Matter, for all of its solidity, is of course not solid at all. There is as much proportionate space in each atom as there is in the solar system. It is truly a miracle that this solid material world that we so thoroughly count on is dependable at all.

Whereas I can live only a few minutes without air, and a few days without water, I can live perhaps a full lunar cycle without food. If I learn to transmute subtle energy into the cells of my body I can exist even longer.

But we must eat. We must eat to nourish our bodies. We must eat to provide the energy that makes our body engines run. We must eat to maintain our temperature. We, as humans belong to a group of animals described as homeothermic, which means we maintain a constant temperature. We are low grade stars. However, as low-grade emitters of energy, we can also modulate some of that energy with our consciousness, thus making these light emissions conscious.

The creation on a good day really is perfect.

You see the birds flying, eating fruit and dropping seeds out their rears into the cosmic dirt. You watch the squirrels climbing trees and eating nuts and burying them in pots and soil all over the yard. The ones they remember, they eat, the ones they forget, sprout to grow into a tree to provide more food. You see a bee on a flower and realize what a gift honey really is. Because our gravitational field punishes big beings, only smaller beings can fly. Small animals can fall many times their height and run away, where men and elephants are flattened. This makes flying cows unlikely.

You have trees and plants that expel the oxygen we need to breath.

You have the seasons, which allow growth in summer, then cause it to flower and bear fruit in the fall, die back into winter, only to regather its spirit and arise in the spring.

You have all the small plants in the oceans that provide the basis for the food chain for an entire population of water creatures.

You have the tides that come and go twice everyday, leaving tide pools where sea birds can fish as if in a barrel.

On a really good day, you see this and more. You feel and know the perfection. You know you are in a wonderful miracle of light and sound and time.

On a bad day though.

You can't find a parking place. The lights on the one-way street don't time out right. They seem to be all red. The car battery is weak. And the bank held a check you deposited for the maximum allowed time, thus causing you account to drop into the red which caused the all-important mortgage check to be returned.

On a bad day, your cosmos seems at war with you own little universe.

And it may well be.

The Creation Is a Creation

I know this sounds tautological.

And I guess it is. But the truth is, the creation is recreating itself every moment. Just as the cells in your body recreate themselves over and over again, so does all matter. The "you that you think you are" that existed 2 years ago is not the "you that you think you are today." Except for a few cells in your long hair or in you nails, everything about you is new. The skin, the cells in the heart, the bones in your spine have all been replaced with new cells that are almost as good as the ones they replaced. If they were exactly as good as the ones they replaced, we could kiss aging goodbye. Instead we become creatures composed of brand new Xeroxed cells that are 99.9% copies of the previous cells. It's that .1% that makes our skin wrinkle and our eyes grow dim.

The Creation is recreating itself every moment in time. So the creation story that we all grew up with is just that, a story. The real story and the real creation is not what happened 6,000 years ago, but what is happening now. The creation did not happen, it is happening now. These words are part of the creation. Your mind reading these words and your being understanding or rejecting them is part of the creation.

It is more like a movie that we suspect.

It is useful, if not a required exercise to see the creation as it is… A Creation.

Imagine that as you watch your lover move from the chair to the bed, that she is an energy form, a ghost moving through a magical subtle fluid that responds to her presence by forming an energy charge or cloud around here energetic form. Imagine that as she moves through time and space, that she is being created by the creation at every moment.

Imagine that everything, every moment is being created at that moment.

Because, you see, that is what is happening.

That realization, truly employed into your operating system, will allow for more degrees of freedom in creative thought which in turn will lead to a more full and pure manifestation.

As J Krishnamurti says, "matter is organized energy." Einstein said it backwards in $E=MC$ squared. But if you divide this equivalency by "C squared" you get Mass = Energy divided by the square of the speed of light. That sounds something close to the same thing to me.

Now, I know it seems a little too much to ask to act as if the creation is a magic field creating itself every moment, especially when you are waiting on tables or waiting in line, but the truth is, if you do act like it, it will respond. If you do not, it will oblige.

This is knowledge.

The Magic Of Time

It's kind of easy to describe the creation. At least it is in very shallow ways. We can speak of atoms and molecules, and gravity and the elements. We talk about how earth circles the sun, and how the sun is part of our galaxy and how our galaxy is one of many in a universe that is seemingly expanding. Our human mind can comprehend how the food we take within our bodies provides us energy to go about our day, and our human mind can comprehend how most things seem to exists and behave.

But time is quite different. It defies description. It defies the brain.

It defies all that we can imagine. We cannot imagine how this all exists "in time."

Einstein postulates that gravity is actually a warp in Space Time. In his model then, Space cannot be separated from Time. His ideas of speed affecting time are well documented in our limited experiments in space travel.

But time does not change just with speed, it changes with age. Or at least the sense of time as our consciousness perceives it changes. I told a young college student one day that the best that I could tell, that "there are 20 years between birth and 10 years of age, and there are actually 10 years between 10 and 20. There are perhaps 7 or 8 years between 20 and 30, and perhaps

5 years in the forties. The fifties may only have about 2 or 3 years in them." As for the 60s, 70s, 80s, and 90s, I cannot say. Perhaps the trend reverses, but I doubt it.

Now this assumes a life that is leaning more and more heavily on routine, the past, and the known.

Khrishnamurti speaks of "Freedom from the Known."

The best way I know to slow down and even stop time is to stop eating. Another, is to suffer great pain and anguish. Or, you can meditate.

Or you can dissolve time itself by the complete submersion into the moment. Powerful moments are timeless.

Thanksgiving

Yesterday afternoon, we celebrated Thanksgiving here in Real de Catorce.

I brought a fine turkey from Whole Foods in Austin Texas. We joked imagining that the turkey was raised in Mexico, slaughtered on the border, packed in California, and sent to Austin, where I returned it home to Mexico to be enjoyed by this truly eclectic bunch. Earlier in the day, I delivered the turkey to the top of the hill where the Swiss wife of my American friend agreed to learn to stuff a turkey. Their house has recently seen a substantial addition with three apartments added. This day was also the birthday of my friend. He is an early Sagittarius.

In Real de Catorce, parties often are composed of people from all over the Planet. This Thanksgiving, we had the indigenous workmen, the Italians, the Swiss, a Mexico City production manager for a Hollywood movie, an art director of a hip newspaper in Texas, Humberto, the Don of the pueblo, and his young teenage daughter Flor.

– The Creation –

There was Mercedes, the south American, and her Italian partner and their children. We had Hector, the photographer, and Bebe, the painter, from Torreone. There was my good friend, Jimmie, the rock and roll businessman, and Luis, the NPR voice of Mexican culture. There were filmmakers, painters, jewelers, a midwife, restauranteurs, and stonemasons.

The food was set on a table on an outdoor patio that affords grand views of the canyon leading to Quemado and Cierra Grande to the south, and a post card picturesque view of the 200 year old church with its grand dome and bell tower to the east. To the west, is a view of the sierra occidentals from a perch generally granted only to eagles.

As the sun moved below the round mountains to the west, the sky turned its usual brilliant red and pink and the high clouds reflected the last light of this day. Knowing that the moon was almost in opposition, I strolled alone to one of the many decks of this modern stone house built in the tradition and manner of the mansions that were built here 200 years ago.

Much of the food was eaten as the moon rose over the Montana de Palos Hablandos.

Then a sense of death came to me.

Moments later, there was a commotion. Flor, the young daughter of Humberto runs up the hill to find her father. There is an emergency.

Humberto leaves with Flor. Several minutes later, Flor returns to retrieve our Swiss hostess who is a nurse and a midwife. There is once again more commotion in the compound.

A small child of a beautiful Swiss couple was found in a water tank.

I begin to wash dishes. My friend from the movies helps me. We clean and clean. Then we get word that they are taking the child to the hospital down the mountain in the city of Matehuala.

There is talk that the child has no heartbeat. As a healer, I am moved to send the energy I can as I wash the plates.

We wait.

Time slows down.

The children sing a birthday song for our host. His wife is almost to the hospital with the 2 year old child. His children lead the song in Spanish. He holds his 8 year old daughter and his 3 year old son in his arms as they all blow out the 12 candles together.

Time is slowing still.

Then Trilce, the American wife of the Italian owner of the corner restaurant grabs me and tells me that my friend is not OK. I turn and find my friend sitting in an erect Mexican chair, with glazed eyes and no consciousness. She is barely breathing.

I bring her to the ground and lower her head and raise her feet.

In a few moments, she is conscious. But her face is pale and her energy and pulse is very low. Earlier, I had remembered the time that I cared for a lady who collapsed on an early morning Southwest Airlines flight. This day, I took charge, directed the stewardesses, and after several minutes, watched as the woman came back to consciousness. For my work on that day, Herb Keleher, the president of Southwest, gave me two tickets to anywhere they flew.

Time now is practically stopped.

I am holding my friends head on the tile floor. She is coming in and out of consciousness. Although I know her well for a new friend, I have not known her long. I do not know of her previous health situations. I only know that she is working long hard hours on the film project that will be moving to this pueblo in a matter of days.

She tries to get up. She lays down again. She tries again. She goes down again. Finally we move to a bed, where she can rest. Then Bebe, the painter and former Ms Mexico, brings a Coca-Cola. It is the energy she needs.

As we are leaving to take the walk to her room with the big windows and view of the pueblo, we are told that the child has a heartbeat. Down below the patio, the vigil has settled around an open fire. Inside, candles have been burning since the news of the trouble began to sink in.

By the time I return to the campfire, the news now is, "there is no heartbeat."

The two year old child dies and the 62 year old man with the 3 year old child celebrates another year in the creation, while his wife must return from a night of human drama deeply knowing and feeling the great loss of the Swiss couple. They have two other children, one almost grown, and another well into grade school. They have now lost their youngest.

For them, time will be different. And their life will never quite be the same. But then, our lives never really are. We just think they are.

But where did the child go?

Now that its young lungs are no longer breathing or being breathed.

Suddenly, the door outside is slammed shut by the wind. I arise and go look. "Hello" I say. "It is just the wind," says Jimmy in the patio below.

Yes, of course, just the Wind.

Now that the child is no longer created by the creation, that form will no longer exist in this Holy Field. That is what happens when we say that someone has died. He will have stopped breathing, he will lose his water, and his carbon will return to the earth.

Is the body just the car that we ride in?

The home we reside in?

The place that we hide in? (Peter Gabriel)

Still, We Grieve.

And the miracle of time will go on. And we will watch it pass by only occasionally marking it with these moments in our lives when it stops or slows to a snail's pace.

The Creator Is the Creation

There is only the Oneness.

The bell tower strikes one.

We know that all matter is energy.

We know that time and space are a continuum.

When we divide the Holy Field, or the Holy Field divides itself,

There is still only the Oneness.

It is like dividing water in the sea.

You may cut it, dice it, name it, manipulate it, and change it.

But the water in the Sea does not care.

It is all still the water.

When the Oneness is individuated.

It is still the Oneness.

Each moment in the Creation is part of the Oneness.

Each individuated part of the Creation is part of the Oneness.

The Spirit is one with the Oneness.

Consciousness is one with the Oneness.

When the Oneness is seemingly divided into the individual,

It is still the Oneness.

When you see with your eyes into the creation,

You see the Oneness from the individuated Oneness.

But the individuated Oneness can only see the reflection of the individuated Oneness.

For there is only the Oneness.

Even when the individuated Oneness is not conscious of the Oneness.

There is only the Oneness.

For the Oneness reveals itself always.

Therefore the Observer is the Observed.

And the Creator is the Creation.

You cannot divide the Sea.

And you cannot divide the Oneness.

And when you divide the Oneness,

There is only the Oneness.

What you see is what you are.

What you are is what you see.

And what you see is the Oneness.

The Practice

Imagine your world as Holy.

Imagine everything in your day Holy.

Salute the morning with reverence and joy.

Greet your first light with the knowledge of your being.

Watch everything as if it is a Holy message from the Holy Field.

Walk with your knowledge, seeing all and knowing all.

Chat with flowers.

Say hello to the dogs.

Listen to the birds gossip.

Pamper that great agave on the corner of your street.

Look into the Oneness through the gate of consciousness.

Stand in the halls of men with the Truth of your knowing.

Speak to the Oneness.

And the Oneness will speak to you.

Laugh at your foolishness.

And cry with those who suffer.

Comfort those who hunger.

And heal those sick that are well.

Make every moment your best.

And every day your first.

Be who you are.

Be wise.

Be understanding.

Be strong.

Be courteous to all.

Open your eyes to creation

And to the Holy Field.

Open your thoughts to the majesty of the mystery.

And to the Sea of our meaning.

Open your heart to this mystery and to the Oneness.

Wash away the myth of your separation.

Find the window to your wishes,

Find the door to your longings,

For there is only the Oneness.

Arise from your dream.

And Awake into the Creation.

And know for certain.

That the Observer is the Observed.

And the Creator is the Creation.

Finale

In the distance I hear drums.

And I hear the footsteps of horses.

And the growlings of men.

They too are of the Oneness.

For there is only the Oneness.

Staying In the Oneness

It is Christmas morning and the air is quite cool and the sun is bright and strong. There was a hard freeze during the night and my water pipes are frozen. My electricity is off, so I write this with the charge of my battery.

The Pueblo is quiet but awakening.

I hear the crow of the rooster.

There is the faint grumble of an engine from somewhere in the distance.

There is a myth that one day after many searchings and much work, that you will be blinded by the Light of Creation and you will forever after be enlightened and reside completely in the mind of God.

Although events of this nature can and do occur with Great Saints, it is a rare and sublime occurrence.

This morning, I read these words of Ramana Maharshi in response to a question regarding the states of mind for a Jnani. He responds:

"How can there be, when the mind itself is dissolved and lost in the Light of Consciousness?

"For the Jnani, all the three states are equally unreal. But the ajnani is unable to comprehend this, because for him the standard of reality is the waking state, whereas for the Jnani the standard of Reality is Reality itself. This Reality of pure Consciousness is eternal by its nature and therefore subsists equally during what you call waking, dreaming, and sleep. To him who is one with that Reality, there is neither the mind nor its three states and, therefore, neither introversion nor extroversion.

His is the ever-waking state, because he is awake to the eternal Self; his is the ever dreaming state, because to him the world is not better than a repeatedly presented phenomenon of dream; his is the ever-sleeping state, because he is at all times without the "body am I" consciousness."

Ramana Maharshi is a 19th and 20th century Indian sage who at the age of seventeen, without the guidance of Guru, attained a profound experience of the "True Self." From this time on, he remained fully absorbed in the Self. After years of silent seclusion, he finally began to reply to the questions put to him by spiritual seekers from all over the world. He wrote nothing and left only very simple instructions to those who would seek him out. Over and over again he would tell them to investigate for themselves, "Who am I."

But most of us cannot and should not retire to a cave or temple.

We must instead nurture our own transcendence even in the desert of the non-reality of Man. For in truth, there is no nonreality. There is only the illusion of nonreality.

At our best, we find ourselves wrestling between impatience and transcendence.

We may live in our Buddha Mind for a moment and then fall back into our Bubba Mind with the slightest provocation.

Yet, if we hold onto the Oneness like a dog to a bone, this Mind of Man will fall away. We will see the Oneness in our friends, in our conversations with them, in the silliness of our minds, in the smallness of our small talk, in the source of our victimizations, in the shallowness of our cravings, in the tombs of our desires.

We will be the Oneness in the quality of our play, and in the depth of our Love. We will be the Oneness in the fullness of our understandings and the in the breath of our care. We will know who we are and where we are and whose time it is.

And like the Jnani, the standard of Reality will be Reality itself.

And the Creator will be the Creation.

And the Observer, the Observed.

You will awake to a place where World has become a Dream.

And the Dream has become Reality.

There is a dog talking to his neighbor.

And two birds are gossiping about the cold night.

And a horse clops up the steep cobblestone street

Thinking of his sore feet.

A rooster crows to his girl friend.

And the power of Man sleeps.

DREAMING

We are all dreaming our lives.

The question is. Whose dream are we dreaming?

Are we dreaming the dream of humankind the victor?

Or are we dreaming the dream of humankind the victim?

There is ample evidence of the connection of dream to reality. There are many, many stories of people who were given numbers while dreaming which ultimately turned out to be big winners in the lotto.

Dreams can tell the future as the story in the Bible recounts. Joseph was wise and he interpreted the dream of Pharaoh in such a way that he brought fame to himself as well as less suffering for the people of Egypt.

Dreams can therefore be instruments that allow us to transcend time. How can this be? Is not time always moving forward?

Dreams can also help us understand the past. This is not so unusual.

There is lucid dreaming where the conscious and sleeping state become connected. Some would argue that this technique is a doorway to knowledge of the self.

There is astral traveling where the dream is actually a realm and the dreamer creates an astral form.

There is dream therapy where the therapist helps the seeker unlock the doors to his subconscious through the study of his dreams.

There are Yogic techniques to control and master your dreams.

There are dream worlds that are open to the sorcerers.

Scientist can now study your dream life and what parts of the brain are being used. We know that dreams occur in certain brain wave frequencies. We know that, thank goodness, we have wiring in our necks that keeps us from acting out our dreams while we are asleep. We know that we need to dream. We know that we may not think that we are dreaming when probably we are not remembering our dreams. We know that dreams are important to mental health.

We know a lot about dreams.

And lots of scientists, therapists, and metaphysicians are studying the content, the form, and the purpose of human dreaming so that we will know a lot more.

And we have the American dream.

My favorite and most repeated dream is not all that special. It's the one where I wake up only to find out that I am still in the dream. I wake up and wake up. When I finally get up in the morning and pinch myself and conclude that I am finally in the real thing, I decide that I am out of the dream.

But am I?

I mentioned to a friend the other day, that as I became more and more in tune with the beat of things, everything seemed to become more and more a dream. If I am thinking something. a word or concept. and it is repeated on the radio station, I think, "aha, the oneness is revealing itself nicely today." Jung had a name for this. I don't.

It's true though. As you begin to see the oneness, the oneness begins to present itself more honestly. That honest presentment is that it is, as the Buddhist would say, an illusion.

So, the reality is the dream. These other dreams are subdreams that work in our other parts of the brain. It seems to me that it might be a bit of a side track to concentrate on the subdreams. Might as well stay on the real thing.

This!

Einstein often said that time and space are a rather persistent illusion.

He wasn't just being cute or flip.

Reality is an illusion. And our brains do a very nice job of keeping it together and rather well explained. But that explanation is just that, an explanation.

We are dreaming our reality.

And I suspect that the morning that I wake up from this dream, that morning will be the morning that my last key stroke has been struck.

Since the oneness is indeed a magic mystery tour of time and space, and all that we know is truly energy, either organized or not, we are all participating in the definition of our reality. Our total combined consciousness is helping the oneness present itself in the form that it is presenting itself.

The pragmatist in you in now saying something like this, "Hey, if I was on an island with no one around, I would still see palm trees and sand and an ocean and a sun and a moon. Those things don't need my consciousness to exist and they don't exist because all of us agree that they do. We may be participating in an energetic time space continuum here, but it is not dependent on me or my race or our consciousness. If we all kill ourselves in a great flame out, there will still be worms in the ground and they will still probably feel the moon and sun. So cut the crap"

OK

What I am saying is this: We are all participating in the definition of our reality. Our total combined consciousness is helping the oneness present itself in the form that it is presenting itself.

We are participating and we are helping.

And you are very right. The sun and the moon will exist without our participation. (probably) The ocean will go on without us. The palm tree will keep on growing. The sand will do what it does.

But your reality is hardly based on these natural things and if it was, it would be quite a differently reality. Our reality is based on the way we shape our reality. From day one, we have agreed with everyone else of our species on all kinds of things.

The Agreements

We have a lot more than the Four Agreements of Miguel Ruiz.

We agree that reality is objective.

We agree that time is unidirectional.

We agree that the laws of physics are in fact laws.

And we agree that we are separate beings.

There are many other agreement of course,

but these are the basics of Our Dream.

Maybe cats think that time can go backwards.

Maybe ants and fish know they are not separate.

– Dreaming –

Maybe dolphins have different laws of physics that fit their watery world.

Maybe the un-programmed child thinks that the reality can be changed. That if you just hope and wish hard enough, if you just wish upon a star, or click your heals, or light a candle, or say it three times, or breath in first, and then blow.

Then, there is this whole issue of the miracle believing adult that goes into the church and prays to St Francis for the healing of their wife, and they believe it can happen and it does happen, and the Doctors don't know why it did. They just say, "every now and then things like this just happen."

So even though, most of us have signed off on these four agreements, most of us kept one finger crossed behind our backs just in case we needed some special attention from God, or whoever it is we think can help us.

I would have added that we all know that we are going to die, but we don't. There are millions of Christians who are hoping for the return of the Lord so they can be whisked off in the so called rapture and they won't even have to die. They are hoping to get the Elijah and the Enoch treatment, who I believe, are the only two humans who got out of this without dying according to the Bible.

Jesus, of course, did die.

He just came back.

So we kind of all agree on these things. And our brains and consciousness is based on these agreements. One of the basic tests for being sane, for being in the human race, is knowing who you are, and being aware of where you are.

Imagine asking Ramana Maharshi who he is after he has spent several years in his cave being totally submerged in the Oneness. Sorry Pal, no you are NOT God. Off you go.

Admittedly, Bhagavan would not make a very good clerk. But he was a terrific saint. And, he did not take the four agreement vow. Or, at least he renounced it at a very young age.

So, let's look at our four agreements.

Reality Is Objective

The truth is, our best scientists know that this is simply not true. The Heisenberg Uncertainty Principle is often misstated to accommodate this notion, but it really only states that you can't observe with complete determinancy. You can't measure the speed of a BB with a train. The train will stop the BB in its tracks. But there are plenty of other things that go on at this level that make the objective reality claim tenuous. How the photon behaves both as a particle and as a wave is a good example.

However, objective reality gets pretty far out when you go the other way and you have to come up with black holes and event horizons in order to describe what is happening in the deep recesses of space.(time)

Einstein's speculations about space/time are equally troubling when it comes to the notion of an objective reality.

How objective is reality for the mosquito who sees heat and not the visible spectrum? What kind of reality does the bat construct?

It is said that the American natives could not see the ships of the Spanish until the Shaman told them that there were ships in the bay.

On a purely mental level, we all know that no two people hardly ever see the same things at a car wreck or a crime scene. That doesn't prove that there was no objective reality, but it does go a long way in showing that position does matter.

– Dreaming –

For example, imagine what happens when a ball is dropped off of a moving object, say a train. To the observer on the train, the ball goes straight down. To the observer on the side, the ball goes in a curve.

Which is the real reality?

From the perspective of the sun, the earth is going in circles, moving around it. From the perspective of the galaxy, the earth and the sun are spinning around each other and moving in a corkscrew motion.

Scale, speed, and position make objectivity a phantom, an illusion. The reality model we use to observe the reality is part of that perspective.

Time lapse photography reveals a world of waving flowers and seeds popping up from the earth like anxious children. Laser photography reveals a completely different world in slowing down of the moment, where a single drop hitting the pavement becomes a major significant event.

What is significant on this scale of the ant or the microbe means little or nothing to us. (Unless of course you have ants in your pants, or a bacterial infection) What is significant to one human life may not be significant to the march of human culture. And the mountain may see human culture as a transitory momentary bother.

To the Cosmos, our small planet is but a speck on a grain of sand on the beach of a world full of beaches, in a universe full of many worlds.

So objective reality is not objective. It is based on position, speed, scale, and time.

Yet it is part of the agreement of the dream.

Time Is Unidirectional

In my mind, I can kind of explain space and things. I can kind of explain how gravity might work or at least model it in some way so that my mind can understand it.

But time is not so easy.

Where did time come from?

The Bible doesn't tell us and most of the creation myths don't speak to it either.

Where does time come from?

And why does it just go forward?

Perhaps it can go backward. Jules Verne imagined it, as have numerous other writers. But few really take it seriously. Time, which we do not understand, which is basically unknowable except as it is, and which is not mentioned in creation myths because it is not knowable, is somehow presumed to behave in a very logical and sequential way. Even though, we all know that sometimes time flies, and sometimes it become pregnant, and it barely moves at all.

The last month of pregnancy is a good example. Ask any mother.

Like I said before, as far as I can tell, there are 20 years in the first 10 years of life, 10 to15 in the next ten, less than 10 years in your twenties, 7 or 8 years in your thirties, 5 years in your forties, 2 or 3 year in you fifties, and after that I don't know, but I'm looking for a turnaround in the trend.

Time therefore is, at least, not objective. For some of us really do have more time than others.

But does it always march forward?

– Dreaming –

How can those who know or sense the future, know? If you believe that someone like Nostradamus or St. John could foretell the future, then time must at least have some characteristics that are not generally accepted or understood.

I am not speaking here of projecting trend lines, I am speaking of the tarot reader telling you that you are going to meet a tall dark stranger in the alley on Friday and you do.

I am speaking of when you know you are going to run into someone and you do.

I am speaking of when you have that feeling that something is going to happen and it does.

Does time leave trails in front of itself?

Do events have event horizons that allow the most sensitive to read and foretell? Or is the moment actually not a moment in the instantaneous sense of the word, but more like a broad line of time that stretches a little into the future and a little into the past?

Time is perhaps our greatest mystery.

But science tells us that if we were thousand of light years in space, and we were watching this planet, we would be watching the past, the way distant past. So, once again, time like perception itself, is based on position, speed, and scale.

Einstein postulated that the faster you go, the slower time will go. Our minds just can't deal with that. Time is constant. Time is linear. Time marches forward. Then, all of sudden, times flies, or maybe it stops.

We've all seen both.

To awaken from the dream, we must set time free. It may go slow, or fast, or backward, or forward.

Besides, it is not so unheard of. If you know the future, you are foretelling. If you remember the past, you are retelling. Time thus moves forward and back in the minds of man, and so therefore, surely also, in the Mind of God.

The Good Ideas Of Physics

Many of the good ideas in physics are not just good ideas, they are in fact the law. But every now and then, things just don't behave. My favorite is water. One law of physics says that things get smaller when they get colder because they become more dense.

Everything pretty much does except water. It starts expanding at about 38 degrees.

Now if water didn't do this, then we would not have ice that floated. We would not have icebergs, and we would not have decent drinks at the bar.

Fish could not live underneath the frozen lakes and things would just not work. So therefore presto, it doesn't obey the law. I liken it to running a red light at 4 in the morning. It is not a law, if it makes no sense. I'm sure the Oneness could see that too.

We don't call this a miracle, but in fact it is.

Personally, I think flying bumblebees are a miracle. They are following no rule of aerodynamics that I know of.

Jesus suspended the rules of physics when he turned the water into wine. He did the same when he walked on the water. He raised the dead, and he gave the blind their sight.

Flying in a 747 would have been a miracle just one hundred years ago if not just a totally preposterous thing. There is no way such a big heavy thing could fly on air. Flying on the electromagnetic field may be just as common in this century once we awaken from the dream which says we can't.

Assembling ourselves at some other point in the matrix we call reality may seem unlikely this century but maybe not in the next. Gravity may become a tendency, not a law. We may find that all the good ideas of physics bend to the stronger forces of mind and soul.

Five hundred years ago, it was a known fact that the sun went around the earth because the earth was after all, the center of the universe. It was a known fact, that human organs could not live in other humans until it was no longer true.

The laws of physics are actually the best ideas we have at the time. If a better idea comes around, we embrace it once we have exhausted ourselves trying to disregard it.

We awaken from the dream when we realize that all is possible in the Oneness and that the light show we enjoy and describe with our tools of description may not always behave as predicted.

We awaken from the dream when we empower our own universe to be as free and alive as we ourselves wish to be.

The Myth Of Separation

Perhaps the strongest sense you have of yourself is well, yourself. We are taught from the time we are babies that we are separate. Sure, we have a family, and a school that we are proud of, and of course, we love our geographic state, that helps define our sense of ourselves, but make no mistake about it, we are taught from day 0 that we are individuals.

WE deify that proposition. We are told that we are the body. We are told that we have a soul that will go to heaven or to hell, so you better be good. We are told that we have Karma that may go with into the next life. We are told that we are Virgos or whatever.

Then, as you begin to investigate the teachings of the great masters, you find that they say quite a different thing. They tell you to find your real self. Who is this real self you think to yourself? They tell you that to find God, you must understand who you really are.

Tat Vam Asi. That am I.

Yes, I get to run my little candy store. And you get to run yours. But we are truly one.

We are one with the creation and we are one with our brethren.

You will not awaken from the dream until the one who dreams finds the dream of separation to be just that.

Allow yourself to be yourself and a strange thing begins to occur. You will find that what is above is below. What is there is here. What is that is this.

Until you are truly alone, you will live in your separation.

You will live in your dream of things and causes and blames.

You will live in your dream of science and physics and naming.

You will live in your dream of personality and power and desire.

You will live in your dream of life and death and sorrow.

But once there is no separation,

It is all an ocean.

– Dreaming –

And love will be your devotion.

And you awake from the dream.

WORKING

What is work anyway?

If you look it up in the dictionary, you will find that the old fashion definition that engineers use is pretty much still the dominant one. That definition says that work is the ability to lift something in a certain amount of time at a certain speed.

The English foot-pound is a measurement of work. It is, as the word implies, the ability to lift one pound one foot in one second. A horsepower is 550 foot lbs per second. That means that you could lift a one-pound object 550 feet in one second, or a 550 pound object one foot in one second. A horsepower is slightly more energy that a Kwh which is a unit of energy based on the non-English system, the Continental system that measures work in ergs.

But this veers off the point.

Work in our society is still measured, when it is measured, as energy.

And that energy is lifting energy, or brute force, not the energy it takes to drive a car, or drive a golf ball, or drive a point home in a poem.

There is no unit of measurement that I know of that measures creativity, or for that matter even skill. It is rather odd that with all of this talk of going from farming and manufacturing to a service economy that we have not developed ways to measure the work inherent in good design, in good aesthetics, and in plain old good common sense.

Sure, if you are a manager of a code writing shop, you can have goals that might equate job performance to lines of code per hour, but these don't actually measure the work or thought process that was involved. If you have a team of skilled masons, you know that you can build a particular wall or room faster and better than the untrained crew can. But you will have to pay your people for their skill. There is certainly no system of measuring the creative process in architecture, or ad layout, or story telling. Yet copywriters, architects, and creative managers know quite well that they are working. Some of them work really hard deep into many a night and well into the best of a planned weekend.

So what is work?

For many of us, far too many of us, work is simply a process by which we gain power. Some of us have to wait tables to get more power, while others must dig ditches, and some play golf, making contacts that will bring that big account into the company. The variety of work in the cultural landscape is truly dazzling.

In developing countries, you seem to have three jobs, the makers, the drivers, and the sellers. More men do the first two, and more women do the selling.

What kind of work is it when a sales clerk sits in a tiny tienda all day and collects money perhaps 20 times during the whole day for the wares they have displayed in their tiny shop that sits neatly next to the Cathedral?

Then again, what kind of work is it when a highly educated lawyer is paid to watch a regulatory process for two weeks billing his client perhaps 15,000 dollars as he monitors the process, perhaps shaping it in some way beneficial to this client?

What kind of work is it when a policeman drives his patrol car down a street doing nothing more than showing his presence to provide a sense of security? How do we measure the worth of the good policeman who is respected versus the cop who is famous for giving out parking tickets, all the while damaging the image of his force and his profession?

Not that long ago, it was easy to define work.

We had people who grew our food and they worked in the fields. We called them farmers. Just 200 years ago, 90 % of the people did exactly that. They grew or raised food. Another small 5% built our houses and maintained our buildings, ships, and shops. Another tiny fraction was composed of merchants. Another small percentage were warriors. The small fraction that remained were simply characterized as noblemen.

In today's developed economy, it only takes 1% of the people to feed the other 99%. And we only need a similar fraction to build, and repair everything else. Even when you include our cars, which do help us keep busy, they only employ a small percent of the population.

In an earlier book, called Lightland, I question the wisdom of supposing that everyone must be working. I question it primarily because, in an advanced robotic society, there really should not be that much work to do.

Advanced programs should run legions of robotic devices requiring only a minimum of human guidance and surveillance.

Instead of moving in the natural direction that our technology is taking us, the ideologues from the flat earth center of the universe club, continue to foster the myth that we must work in order to gain the power to purchase our food, our homes, and our wide screen TV.

We therefore create rather silly jobs that machines would actually do better. As I said in Lightland, we don't really need most of the jobs that we have in a truly advanced economy.

This kind of working, this rather silly and pointless exercise in human bondage and control is not the Work that is referred to in this chapter. And although this certainly merits a full discussion which reveals and dissects the injustice and inequality inherent in such a system, that is not the purpose of this book.

I want to speak to the other definition of work.

That definition is closer to "operating well."

"Is it working?"

"Yes. It's been running now for about an hour and it seems to be working fine."

That is the Working I want to discuss.

Operating Well

Working in the creation and in the oneness is not working at your job. Working at your job is a tiny subset of your real work, your real worth, and your real value to the community and to the creation as a whole. To be Working, is to be functioning properly, to be operating well.

For an individuated aspect of the creation to be working well with the creation, it must be aware of its place, potential, and purpose. To be Working in the creation, the individuated one knows that it is, and that it is in the oneness. It knows that it cannot be divided from this sea of meaning, yet it knows that for all appearances, it is.

To be Operating Well, the one awakens from the dreams of men and it awakens to the reality of the creation and the Holy Field.

No matter if you have a job or not, you can and should be Working.

For there is so much Work to do.

Each of us is an operator in the Holy Field. We are capable of making our world better or worse every moment we breath. Moment by moment, by the choices we make, by the attitude that we hold, by the mood we embrace, by the belief system we employ, we are making changes and waves in the creation.

We have been given an enormous gift and responsibility.

Over and over, again and again, moment after moment, day after day, we have a seemingly infinite amount of choices in regard to our conduct and our effect on the world around us. If we greet our children with a hug and a kiss, and our lovers with a strong embrace, we have made the world stronger, and brighter, and more loved.

But these are the easy ones.

Try loving the cop that stops you for speeding. (And yes, of course you were speeding).

Try loving the clerk that won't give you the permiso you need because your car title has a lien holder and they have a new policy that requires all lien holders to give their permission for a vehicle to be allowed into their country.

Try loving the drunk who offends you and tries successfully to be obnoxious to the point that you either have to leave, change his attitude, reform the reality, or call the police.

Try loving the daughter who won't learn from her mistakes or the son who won't stop blaming others for his own shortcomings. Try loving the girl friend that lies, or the wife that found another man. Try loving the man who steals from the needy so he can keep his wife in diamonds and his lovers in new cars. Try loving the banker that screwed your house loan up after assuring you that all was well, because he didn't know his job well.

And, sometimes, operating well does not just mean being all lovey duvey. It may require a quick and stern response. Operating well means being your very best, as much as you and your friends can stand it. You will be called upon to be decisive. You will be called upon to act quickly and confidently. You will be challenged. At times, you will do great. Sometimes you will fall short of your potential. That will still be operating well.

Operating well means behaving at all times as if it matters.

Because, it probably does matter.

Operating well means listening and watching the signs that appear before your eyes and to your other special sensory abilities. It means talking to the creation in all of its manifestations. It means listening to the rocks and to the trees. It means touching the earth while you embrace the sky. It means walking in balance down a razor's edge of impersonal discernment and wizened experience. It means learning from the hurt and forgiving the hurter.

Operating well means meeting every day with as much enthusiasm as you can muster, and sometimes that may not be very much, for everything that moves must find rest.

When I am at work, I often bump into people on the elevator. Invariably, the conversation goes something like this:

How's it going?

Well, it almost Friday.

Or there is the other one,

How's it going?

Well, its almost five.

Almost Five, Almost Friday

These people are not operating to the fullest, yet they are pretty good folks. They are paying their bills, sending the kids to school, and doing their job responsibly for the most part.

Paradoxically, I find that many of these folks are quite happy in their jobs. They just think that they are supposed to not like it, and so therefore they act as if they can't wait to get home to those horrible TV shows, that dirty laundry, or that grass that needs to be cut.

One day while in the elevator, I tried a different response.

"So, I don't get it, if it's Friday, then it's almost Monday, and we all really hate Mondays. So how do we make sense of that?

The question at least caused a good inward moment. And the guy in the elevator knew exactly what I was talking about. He knew that each day is a gift, that each moment is a spectacle, and that each hour that we are alive is our own miracle.

This brings up the other aspect of operating well, and that is always being mindful of the magic of the moment and of the power of the present. We cannot operate in the fullness of our being if we do not see each moment for what it is, the living moment from which all history is written and all future is fomented.

So even as we do our job so that we may have enough power to buy our food and supply our family with a house and cars, we need to know that that this is not our work. Our work goes on after five and it goes right through the weekend. Our work is, as the saying goes, never ending.

And that is as it should be.

For those who choose not to live their lives fully, it will always be almost Friday and almost Five. Like a mirage, those who wait for Friday will find that it never really gets here. Those who wait for their happiness to begin after five, will likely not find the happiness they seek, either in the bottle, in the cigarette, or in the programming on their new wide screen TV.

The Practice

Driving down to my mountain home this trip, I tried something new. And, in all fairness, it worked pretty well. I had mentioned in another piece that I would be with my family during the trip, knowing full well that I was traveling alone.

I am always pleasant to strangers. I almost always show them the dignity we all deserve. But this time, I took a different stance. I treated everyone as if they were family. Clearly, in the largest sense of the word, they are. Whether we believe we came from Aunt Lucy or not, we are still very related. Our DNA is pretty dang close.

When I came to the detour around the new tunnel, and I approached the construction worker who was flapping his torn flag at every driver, I slowed, perhaps a little more than normal, and then I waved as if he was my cousin. He was a little surprised, but then he waved back. And as my wave was not a "I don't know who you are wave", his wave back also was not a "I don't know who you are wave." It was weird. We waved as if we had gone to high school together.

I tried it on the flag-man on the other side of the detour. And it worked just as well. We waved heartily at each other. All day I acted as if I was traveling with my family. When the truck driver pulled in front of me on the hill going 20 miles an hour, I gave him the slack I would give my cousin. When I bought a pair of cheap sunglasses, I put them on and asked the clerk what she thought. But I didn't just throw out the words, I asked her what she thought, as if she was my sister. And she responded (in Spanish) that they looked pretty good.

Now that I think about it, I suspect a real sister would have been much more critical and crabby.

This whole trip I have used the same method. It has worked well, except for the man who clearly did not feel close enough to give me the manly bear hug that is often used between good amigos. I found myself moving in without the complementary movement from my greetee. I can handle that kind of embarrassment; besides, it taught me a few subtleties about body movement and non-verbal communication. If I had been real conscious, I would have picked it up earlier and in time to avoid the little humiliation that followed.

It seems that in truth, the more you let your light shine, the more likely other lights will perk up too. Brightness begets brightness.

There are many other ways to view your interactions with your fellow human family, but this one seemed to work for me. Perhaps you have the knack of politician and you are genuinely glad to see and visit with anyone.

I am not that way.

But whatever system or natural talent you may have, there are certain things that we know are symptomatic of not operating well. We rarely win friends and influence people positively though pride and greed. We rarely make things better with anger and I know of no one who has ever impressed me with their ability to hate. We rarely solve a problem with force of violence. We rarely make the world a better place by hanging on the horn as we rush through the reddening yellow light. In short, only on rare occasion is acting like an ass hole an example of operating well.

We are not operating well when we are so self centered that we dismiss our own impact on others. It is odd that supremely self centered personalities tend to marginalize themselves by not giving their actions the full weight they deserve.

It's a "I'm so important I don't matter" kind of thing.

You don't have to be mystic that is firmly rooted in the second attention to be operating well. You don't have to be anything but your best. And you generally know down deep what that is. That is all that is required, and that is a very tall order. Most of us learned how to operate well in the world way back in kindergarden. And most of us know better.

We know how to operate well.

But why do we even care if we working, or operating well, or acting morally?

Or more accurately, why are so many of us not doing this?

Why are so many content with their consciousness the way it is?

Why are so many not the least bit interested in evolution in themselves or for our race?

Why are you reading?

Why am I writing?

Why are we doing this?

These are not questions that are posed because their answers are noteworthy. The answers to these questions are absolutely essential to understanding ourselves, our world, and our potential.

Why Are We Doing This?

When I was in my twenties, I became aware of something called transcendental consciousness. Perhaps it was the right combination of the Beatles, Walt Whitman, and Buck's book on Cosmic Consciousness.

I wanted to be realized!

I wanted to be a realized Being.

For several years, I read everything I could get my hands on.

— Working —

And it wasn't too long before I came to my own conclusion on that whole process. I became hip to the big lie. It was clear that it was very difficult to convey the truth with abstractions that issue from of our subject object mind platform. I even wrote a song about it. It was called the Model is the Madness.

The words went,

Well I'm not trying to say

What to say,

And I'm not trying to say

What to do,

Seems that when you climb a rung

Everything falls,

And when you are asleep,

You are really awake.

And when you think you're moving

You're approaching the static,

Ain't it funny?

I'm estatic.

The Model is the Madness the Model is Madness the Model is the Madness the Model.

Simply stated, a picture of a tree is not the tree. It will not shade you. The picture is an abstraction of the tree.

So, it is very difficult to communicate new mind shapes and new symbols with the current state of programming. Our current program is simply not capable of expressing these other mind shapes.

Perhaps, that is one reason that art, poetry, and song can be more effective at communicating transcendental states than well formed words. Using the magic of music, other doors to the soul can be opened. Using poetry, new metaphors help replace worn out metaphors that are no longer valid or healthy.

The artist, poet, and songwriter among us have a unique skill set for moving human consciousness forward.

So, let's go back to the question. Why do we want to be better? If in fact we do? And why do some of us not really care about it that much? Pulling no punches here, they could give a shit. Now, stay with me now, because this is critical. If I want to be a realized being because I think that somehow that gives me a leg up on every one else, then I need to investigate that.

The Oneness is not going to be fooled that easily.

If I want to be a better, more moral, more spiritual person because it suits my view of myself, then am I really any better than the man who pursues wealth and power with pointed fury? Am I really any different from the ones who could care less?

And if you are reading this because you want to be better, smarter, more moral, more spiritual, and just generally more with it, then the same goes for you too.

If the groundwork beneath our motives is no different than the groundwork of Julius Caesar's desire to conquer and unite, what good can become of this?

Building on this mind is like placing a new operating system on top of an old system.

If we do, it will just become a more complicated version of the original operating system. It may be superior in many ways to the older system, but it can never be profoundly and fundamentally different.

Yet, I would argue that it is probably true that using this present operating system to its maximum capability is certainly not a bad thing, and it might be able to point in the general direction of the next operating system that we can build and refine.

And, there is this force of compassion.

Reducing suffering and injustice is always a good idea even if the results are not what you expected. The compassionate heart works with any operating system.

And that is why some of us care.

And some do not.

If you believe that this operating system is headed in the wrong direction and that the results that it may produce may not be what we want . . .

And if you believe that there is work to do that can help alleviate or mitigate suffering and injustice . . .

And if you believe that the only way to really change the events that lay ahead is with a substantial increase of more advanced mind shapes that can interfere with the dominant dream, then it seems understandable that we would and should employ our best to make that happen.

A compassionate heart trumps an underdeveloped sense of self.

And I think that answers the question of motive. At least for me it does.

For me then, it is clear that we must practically throw everything out. The whole erected operating system is entirely too archaic to handle the knowledge we are accumulating with our increasingly expanded differentiation abilities.

First, we throw out the concepts that divide us. That means we must redefine our beliefs regarding nationalism, racism, class, everything. We throw all of those concepts that shackle us to our archaic understandings out the door.

We forget about capitalism, free enterprise, communism, socialism, Darwinism, Creationism, full employment, a healthy economy, democracy, and freedom as most people imagine it. Those things we hold dear, we discard because, in truth, many are nothing more than a thin façade of legitimacy wrapped around a great mass of archaic power.

We find a new freedom in that unyoking of the mind and the human spirit.

We radically change our sense of self and thus unlock the great hidden potential of humankind.

We unyoke ourselves from the fear of death, and we learn to walk with death and we make it our ally.

We sharpen our minds to be sharp strong swords that can cut what we observe so expertly and cleanly that the whole is not lost in the process of observing.

We observe without judging, for we are just judging ourselves.

We focus on feeding, and providing, and caring,

For the Creation as we know it.

We focus on beauty.

And we explore with a liberated mind

WE explore into the Oneness.

This is Work

This is Working

This is operating well.

With a new operating system.

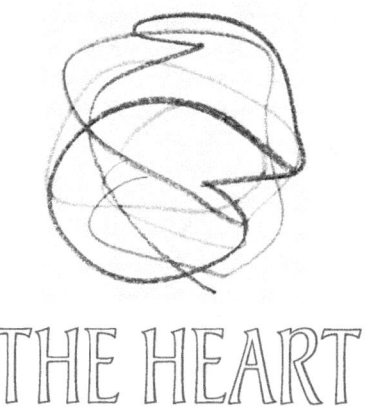

THE HEART

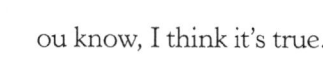You know, I think it's true.

We wouldn't need a brain if we all had heart.

But alas, many, as the story goes, are like the tin woodsman.

Oh, of course, we have hearts.

I mean Heart.

Many of us live in our minds.

And some of us live in our hearts.

There is some recent research that indicates that the organ of the heart is full of the same nerves and much of the same material as is found in the brain. So the heart not only pumps the blood and the oxygen to all of the rest of our cells, it also thinks or feels.

There is some evidence of this in heart transplants.

It seems that recipients of heart transplants begin to take on certain characteristics of the former owner of the heart. This is rather startling news for the "everything is only what it is" set.

If someone from the loopy corner of life had suggested that such a thing might occur, as they were considering which heart to transplant, they would be thoroughly and completely marginalized for such silly, unscientific thought.

Apparently, it is scientific.

So, the heart and the brain both have thinking and feeling capabilities.

We all know this. It just took a while for the smart guys to know it.

The Awakened Heart

I'm not quite sure why our hearts need to be awakened. Did we allow them to go to sleep as we became small adults, or does our heart lie dormant awaiting its time to blossom?

Either way, most of us know or have experienced a glimpse of the awakened heart. Perhaps it was the death of our father, or the death of a sister or brother. Perhaps it was your own near death.

Perhaps it was not death that awakened your heart.

Perhaps it was life as found in the new born babe.

Perhaps it was the care that was given to you as a child.

Perhaps it was the love of your partner.

Perhaps it was the love that you gave,

when you wanted to hurt because you were hurt.

And you saw your error.

And your heart awakened.

I think we all know that the love we have for our partners is often not love.

I think we understand that the love we have for our children and parents is often not real love. I think we comprehend that the love we have for our friends and neighbors is generally not love.

Many times our love is self serving.

– The Heart –

Many times our love is charm and charisma.

Many times our love is just a good idea.

The love of the awakened heart is something different. The awakened heart steps in when you are getting ready to strike back at someone who has offended you and it says to you, "if you take this, you can help stop it, but if you return it with your own vitality, you will cause more hurt."

In fact, as I wrote these words, I received an email from a writer who is currently enjoying a lot of popularity with his book, The Long Emergency. It is about the end of Oil and the end of our reckless consumer ways and idiotic way we in the West have designed our post-war, car-centric cities.

We had been visiting about his drive to his summer home and my drive to this mountain home away from home.

We very much think in very similar ways. I had actually e mailed him to apologize for subconsciously using one of his story titles in one that I just written. He wrote back that it was probably synchronicity.

I wrote back with a quick note. He responded with another quick chatty note.

I responded to his little chit chat e mail with a more lengthy, but arguably witty and entertaining piece about the Huicholes and where the sun was born, and isn't it good that a solar guy like me lives next to where the sun was born.

The next e mail I got back from him was a plea to not engage in chit chat, because his time is valuable to him. And arguably to us too.

He hurt my feelings, my ego, my sense of importance.

So I sent back a message that said,

"Are you bipolar or what?

You chit chat with me and then I chit chat with you and I get this piece of crap e mail that says please don't chit chat."

He responded back to that, with "don't be mad."

And I responded back that "I will get over it." That it was a modest dent in my tranquility. But this tranquility that I so value was entirely too dependent on my sense of self importance. I could have just as easily complimented the man for his diligence in actually answering my e mail in the first place...

By the way, he probably is "you know what."

I have experience.

The love of the awakened heart is not the love of your self importance.

The love of the awakened heart is not the love of sentimentality.

The love of the awakened heart is not the love of your own goodness, of your own forthrightness, of your respect for others, or of your honor to your own words and commitments.

The love of the awakened heart is not the love of the breast of a woman — or the loins of a man.

We know.

The love of the awakened heart is the natural state of a child lost in the mystery of the sea of our knowing.

The love of the awakened heart is the love for a woman who left you for another man.

It is the love for a man who left you without an education or youth.

It is the love for a father who touched you and violated you.

It is the love for a mother who had no time for you.

It is the love for a boss who treats you like a simpleton.

It is the love of a red light on a busy day.

It is the love of a bee sting.

It is the love of a blue sky that turns stormy,

and rains on your parade and party.

It is the love of your enemy,

Who would kill you if he had a chance.

As Jesus said in the 5th chapter of Matthew,

"Ye have heard that it hath been said, Thou shalt love thy neighbour, and hate thine enemy.

But I say unto you, love your enemies, bless them that curse you, do good to them that hate you, and pray for them which despitefully use you, and persecute you;

For if ye love them which love you, what reward have ye?

Do not even the publicans the same?"

The Compassionate Heart

When I go into the Cathedral that silver built in this high desert mountain town of 9,000 feet, I speak with the woman who sells the candles and the milagros, I say hello to Arturo who sells me the box cameras I love to use, and I pass by the others who vend their remarkably kitchy wares on the steps of this holy healing place.

From all over northern Mexico, perigrinos come to this place. Often, they even walk. And when they get to the old wooden floors of the church, they walk on their knees to the altar of St. Francis, who sits on a throne in his neon lit glassed-in altar. Always, there are perigrinos kneeling and praying to their patron saint.

When I am here, I go to the church every morning and I walk past St Francis but only after kneeling briefly and genuflecting. I prefer to go to the back, off to the right to the room of the "Particularly Tortured Jesus." Here, is a Jesus that was crucified.

Each time I walk in, I say to myself, "Jesus, what have they done to you? You look awful." The blood is dripping off his kneecaps, out of his armpits, and from the gaping wound of the spear in his right side. The nails in his feet and hands look more like railroad spikes and the blood is oozing from these wounds almost in front of my eyes.

He has a nice wrap on with a golden sash, but his hair, yes he has real hair, is a mess with that crown of thorns and all that blood dropping down on his almost serene face.

This is a real torture.

In Mexico, perhaps like no other place, the compassionate heart is truly alive and dripping. The Patron Saint of the Country is our Lady of Guadalupe. Her heart is with the poor people and those who suffer.

— The Heart —

Her first appearance was with Juan Diego. Here is the story.

"On December 9, 1531, a native Mexican named Juan Diego rose before dawn to walk fifteen miles to daily Mass in what is now Mexico City. Juan lived a simple life as a weaver, farmer, and laborer. That morning, as Juan passed Tepeyac Hill, he heard music and saw a glowing cloud encircled by a rainbow.

"A woman's voice called him to the top of the hill. There he saw a beautiful young woman dressed like an Aztec princess. She said she was the Virgin Mary and asked Juan to tell the bishop to build a church on that site. She said, "I vividly desire that a church be built on this site, so that in it I can be present and give my love, compassion, help, and defense, for I am your most devoted mother . . . to hear your laments and to remedy all your miseries, pains, and sufferings."

"The bishop was kind but skeptical. He asked Juan to bring proof of the Lady's identity. Before Juan could go back to the Lady, he found out his uncle was dying. Hurrying to get a priest, Juan missed his meeting with the Lady. The Lady, however, met him on his path and told him that his uncle had been cured.

"She then told Juan to climb to the top of the hill where they first met. Juan was shocked to find flowers growing in the frozen soil. He gathered them in his cloak and took them at once to the bishop.

"Juan told the bishop what had happened and opened his cloak. The flowers that fell to the ground were Castilian roses (which were not grown in Mexico). But the bishop's eyes were on the glowing image of the Lady imprinted inside Juan's cloak.

"Soon after, a church was built on the site where our Lady appeared, and thousands converted to Christianity. Our Lady of Guadalupe was declared the patroness of the Americas."

And She hangs above me now in a nice piece of kitch that reveals Our Lady or St Francis depending on the viewing angle.

The Compassionate Heart is the heart that feels suffering, whether it is yours or not.

The Compassionate Heart is the understanding of the Holy Mother who feels the life in all things, and in all of creation.

The Compassionate Heart is the opened heart of the truth of the creation and the principles of giving and love.

The Compassionate Heart are the tears that stream down your face when you see that beautiful rainbow that arcs over the dome of the church, or a sunset that caresses the sky and its brethren clouds with colors that would even make the rainbow jealous.

The Compassionate Heart hears the cries of those in Africa and it sees the loss of true soul in the West.

It does not think, it feels.

For it is one with life.

Chitta

In Buddhism, there is this word "chitta." This excerpt comes from Hans Gruenig.

"In the Buddhist pedagogical tradition, the noble, liberating emotional qualities are sometimes lumped together under the term "compassion."

Walpola Rahula writes:

"According to Buddhism, for a man to be perfect there are two qualities that he should develop equally: compassion (karuna) on one side, and wisdom (panna) on the other. Here compassion represents love, charity, kindness, tolerance and such noble qualities on the emotional side, or qualities of the heart, while wisdom would stand for the intellectual side or the qualities of the mind. If one develops only the emotional neglecting the intellectual, one may become a good-hearted fool; while to develop only the intellectual side neglecting the emotional may turn one into a hard-hearted intellect without feeling for others. Therefore, to be perfect one has to develop both equally. That is the aim of the Buddhist way of life: in it wisdom and compassion are inseparably linked together."

Wisdom and compassion are two sides of the same golden coin of enlightenment just as heart and mind are two sides of heartmind.

Sharon Salzburg notes:

"In Buddhism there is one word for mind and heart: Chitta.

Chitta refers not just to thoughts and emotions in the narrow sense of arising from the brain, but also to the whole range of consciousness, vast and unimpeded."

While moral behavior, meditation, and the cultivation of wisdom can propel us towards enlightenment, skillful invocation and emulation of enlightened mental qualities can potentially 'prime the pump' and speed up the purification process, allowing us to benefit more widely more quickly.

For this reason, many meditative traditions offer various ways to invoke those enlightened mental qualities — qualities such as metta (loving-kindness) — even before it occurs that they arise spontaneously and ubiquitously within one's heartmind."

The Heartmind

I have always thought that the cross was literally that.

A cross.

It is both a metaphor and the key.

As you cross yourself in the Catholic tradition, you go up to down, to left to right, then to the heart and back. (there are derivations)

For me, this crossing is the cross of the mind and the heart.

The Heartmind.

The Heartmind is the blend of compassion and wisdom.

It is the simultaneous mutual identification with life and this world with consciousness and the creation of mind.

It is the point where a new organ develops or becomes activated.

It is the Compassionate Heart and Awakened Consciousness combining into a new oneness of life and light, of the individuated and the whole, of intuition and intellect, of in and out, and of me and you.

The Heartmind becomes the new house organ of oneness and clarity.

This is not a new operating system.

And it not even new hardware.

But it is hardware that must be recognized by the system.

You may need some new drivers.

Truth Is a Pathless Land

"Man cannot come to it through any organization, through any creed, through any dogma, priest or ritual, not through any philosophic knowledge or psychological technique.

"He has to find it through the mirror of relationship, through the understanding of the contents of his own mind, through observation and not through intellectual analysis or introspective dissection.

"Man has built in himself images as a fence of security—religious, political, personal. These manifest as symbols, ideas, beliefs. The burden of these images dominates man's thinking, his relationships, and his daily life. These images are the causes of our problems for they divide man from man.

"His perception of life is shaped by the concepts already established in his mind. The content of his consciousness is his entire existence. This content is common to all humanity. The individuality is the name, the form and superficial culture he acquires from tradition and environment.

"The uniqueness of man does not lie in the superficial but in complete freedom from the content of his consciousness, which is common to all mankind.

"So he is not an individual.

"Freedom is not a reaction; freedom is not a choice.

"It is man's pretense that because he has choice he is free. Freedom is pure observation without direction, without fear of punishment and reward. Freedom is without motive; freedom is not at the end of the evolution of man but lies in the first step of his existence. In observation one begins to discover the lack of freedom. Freedom is found in the choiceless awareness of our daily existence and activity.

"Thought is time.

"Thought is born of experience and knowledge, which are inseparable from time and the past. Time is the psychological enemy of man. Our action is based on knowledge and therefore time, so man is always a slave to the past. Thought is ever-limited and so we live in constant conflict and struggle.

"There is no psychological evolution.

"When man becomes aware of the movement of his own thoughts, he will see the division between the thinker and thought, the observer and the observed, the experiencer and the experience.

"He will discover that this division is an illusion. Then only is there pure observation which is insight without any shadow of the past or of time.

"This timeless insight brings about a deep, radical mutation in the mind.

"Total negation is the essence of the positive. When there is negation of all those things that thought has brought about psychologically, only then is there love, which is compassion and intelligence."

J. Krishnamurti

The Spontaneous Arising

Maybe you will be getting a massage, or you will be riding in your car into a golden sunset. And you will suddenly find yourself leaking out. You will not be able to contain yourself. You will feel and know the truth of your beingness.

You might feel a tingle that goes up your spine.

You will feel a fullness and a brightness.

Tears will shoot out of your eyes like a firehose.

Everything will be so beautiful.

I mean truly beautiful.

You will marvel at the trees, at the earth, at the sky.

You will see the rain as a miracle.

And the ocean as a great feat.

You will marvel at the incredible notion of time.

You will be spellbound by the greatness of space.

You will suddenly know that you are a spirit in your body.

You will understand that you are your words, your actions, and your works, and as others read and hear your words, and feel your actions, and experience your works, you become extended well beyond the boundary of your skin or the area of your presence.

You will identify with all of life, with the birds, and with the rocks.

You will see yourself in others.

You will feel their pain, because it is your pain.

You will know them.

As you will know yourself.

You have become the Heartmind.

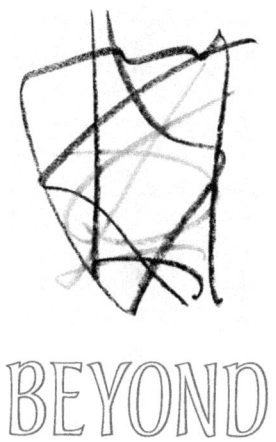

BEYOND

— *Beyond* —

*I*t was a super bowl Sunday when I started this book.

It is a super bowl Sunday again today.

Since the beginning of this effort, there have been perhaps four maybe five super bowls. I have versions of this book on 3 computers. One is dead and gone now. I write today on the one that was used for most of the chapters, but it is not the one I use most days.

I have written several other books during this time and have seen two of them published. Last year alone, I wrote more on my blog than I have written in all of my published print works.

Most of the chapters of this book, with the exception of the first chapter, the chapter on the sea, and this last chapter, were written from my mountain home in Mexico.

My home there affords a vast panorama of strong round mountains and thick blue sky with a rich culture of Huicholes, other indigenous peoples, and an international component of artists, photographers, shopkeepers, and jewelers.

My house sits right on the park, where I watch the children play late in the day, and glance down at the old men who sweep up in the early mornings.

It would be a lie if I said I didn't know who was playing in today's game.

I do.

I just don't know how they got there, who the players are, or who is the better team.

That would not have been true in years past.

There have been times when I watched the pro season with the joy and enthusiasm of any other fan. There have been times when I would have my week somewhat effected by the loss of my team on Sunday.

There have been times when it was very important to me to be part of the sports entertainment world and all that is has to offer.

But, I have gone beyond that time.

Now, when I turn on the TV and watch this great pastime, I find myself simply inundated with a culture that is just not right for me anymore.

The cars, the beer, and the other products that are promoted and elevated by their presence on this day's broadcast simply have no place in my world these days.

Of course, I still have a car and I still buy computers.

But very little advertising gets into my world.

I have gone beyond that.

Perhaps you too have gone beyond television and the culture that it weaves on our minds and our souls.

The Process

Some part of me, perhaps a very large part of me, was hoping that as I wrote this book, that I would get to this last chapter and be able to write from the standpoint of someone who is wise, someone you would want to be like, someone with whom you would place great trust.

That part of me wants to be a great teacher, a great sage, a great something or another. That part of me wants you to read

– Beyond –

these words and to be amazed and entertained. That part of me wants to be on some TV show or on tour in front of a bunch of admiring fans.

That part of me wants to be famous and to be known in the streets where I reside.

Being some kind of mixture of prophet, poet, and creative scientist, that part of me was somehow willing and anxious to share what I know and to perhaps tell how you might use that knowledge to further your own life and conscious development.

That part of me is still here.

But another part is growing now.

Now, another part has found root in the golden consciousness of our own light.

That part knows the sillyness of language, the vanity of religion, and the sublime lie of the individuated consciousness.

That part knows the mystery of space and the wonder of time.

But, it knows not the meaning of the creation or the nature of its dance.

That part, that growing part, sees my daughter who has cancer, my ex-wives who left me, that trail of stupidity in my wake of days, and it sees the perfection of it.

That part, writes today, not as a realized being who has found some cosmic consciousness, but as a being who has lost its edges.

That part, is in no way, a conversion of the old part.

It is a disintegration of that part, which paradoxically, leads to integration.

Our Natural Condition

It was the beginning thesis of this book that transcendence is our natural condition. I state in the first chapter that we all seek transcendence into the moment and into the one through various techniques and processes.

Some of us watch football, some of us go to movies, some of us quiet our minds in yoga, some run, many use the drug or drugs of their choice.

It is our Natural Condition to be transcendent, for we are all transcendent beings.

And, it takes a lot of meta-programming to convince us that we are not.

Yet, I know that each of us are light emiting creatures with energy forms that are capable of the most wondrous acts of creation and purpose.

Through our understanding of the door of the eternal moment, and of the fields of light, we can find our true nature.

Through our alignment within the cosmos and the sky itself, we can find our true place.

Through our understanding of the power of our mind and our process of naming, we can judiciously apply our powers of judgment and discrimination.

As creatures of light, we realize ourselves capable only of seeing ourselves, as long as we hold onto a self that is limited and disintegrated.

Through the embracing of our true being, we can find our natural state and we then can believe a new dream, a new creation, and a new state of being.

For when we do, we arrive to a bright world of spirit, of the soul, and of the creation itself, where, with our own unique dream, we begin to work in this world of light from the center of our newly realized and connected heartmind.

We find in the fullness of this state, that the great mystery of time itself wraps it shimmering blanket into each moment and over each day.

We know that to try to pierce its shroud, is as foolish as writing on the wind.

But mostly, when we look deeper and deeper into ourselves.

We begin to see and to feel that there is no there, there.

We are a bundle of light and energy, thoughts and memory.

We are a mystery within a miracle.

And we are truly one.

Beyond.

About the Author

Although Michael J. Osborne is best known for his books and work in renewable energy and in climate change, his primary interest and expertise is Energy.

His first book, *Lightland, Climate Change and the Human Potential*, is a blueprint for our civilization to move forward as an ethical and prosperous global community. His second book, *Silver in the Mine*, was written to provide a formal road map for the City of Austin, as it moves towards becoming a sustainable city.

Now, in this new work, Michael J. Osborne lays out a blueprint for a new operating system for the subject/object consciousness that we humans presently use in making our way through the mystery that we call existence.

Through his work at a major electric utility and through his blog, *earthfamilyalpha*, Michael works everyday to transform the way we make our energy, and to transform the way in which we allow our own energy to be shaped and channeled by the institutional and epistemological relics of our past.

In *Beyond Light and Dark* the author shows us the mystery of these energy patterns that we have created, and he helps us unravel them.

Mr. Osborne lives in Austin, Texas and in Real de Catorce, Mexico with his partner and family and friends. He may be contacted through his blog.

About the Artist

The fourteen drawings in this book were created by Charlie Tomorrow over the course of many months after reading parts of the original manuscript of Beyond.

Charlie is a an abstract painter and musician who has lived in Mexico in Real de Catorce for the last 15 years with his artist partner Cora Vann. Born in Germany, Charlie and Cora traveled the world before making their home and studio in their high mountain pueblo in the Sierra Catorce.

Charlie and Cora's work can be found at Gallery Vega M57, and in Galleries in Mexico City, Zacatecas, Austin, Texas, and in Colon, Germany. (http://vega.m57.googlepages.com/)

Charlie's work is intended to create the maximum peace and freedom in the viewer and within himself, and to move the observer towards a peace and freedom that is ultimately found in the One.

All the rest is motion.

www.ingramcontent.com/pod-product-compliance
Lightning Source LLC
Chambersburg PA
CBHW051427290426
44109CB00016B/1463